FINANCIAL TIMES MANAGEMENT

Knowledge   Skills   Understanding

Financial Times Management is a new business created to deliver the knowledge, skills and understanding that will enable students, managers and organisations to achieve their ambitions, whatever their needs, wherever they are.

Financial Times Pitman Publishing, part of Financial Times Management, is the leading publisher of books for practitioners and students in business and finance, bringing cutting-edge thinking and best practice to a global market.

To find out more about Financial Times Management and Financial Times Pitman Publishing, visit our website at:

www.ftmanagement.com

Managing technological innovation

Fourth edition

Brian C. Twiss

FINANCIAL TIMES
PITMAN PUBLISHING

LONDON · HONG KONG · JOHANNESBURG
MELBOURNE · SINGAPORE · WASHINGTON DC

FINANCIAL TIMES MANAGEMENT
128 Long Acre, London WC2E 9AN
Tel: +44 (0)171 447 2000
Fax: +44 (0)171 240 5771
Website: www.ftmanagement.com

A Division of Financial Times Professional Limited

First published in Great Britain in 1974
Fourth edition published 1992

© Pearson Professional Limited 1992

ISBN 0 273 03795 1

British Library Cataloguing in Publication Data
A CIP catalogue record of this book can be obtained from the British Library

10 9 8 7 6 5 4

Typeset by Avocet Typeset, Brill, Aylesbury, Bucks
Printed and bound in Great Britain by Bell and Bain Ltd, Glasgow

The Publishers' policy is to use paper manufactured from sustainable forests.

Contents

List of figures

List of tables

Preface

Specialization of knowledge, division of labour and detailed job specifications are a common managerial response to the complexities of the modern world. Attention is focused on the minutiae of managerial problems and techniques for dealing with them. Yet when we examine the successes and failures of business we can usually trace their origins to management's performance in respect of a few 'big' decisions. Often these were decisions about an investment in technology or a technological innovation. Of course, it is desirable that all decisions should be sound. But the effect of one poor decision which affects the whole business cannot be mitigated by a host of correct minor ones.

Having focused the individual manager's attention on a limited range of parochial problems it then becomes difficult to bind his work into a system or overall plan giving purpose and coherence to the contributions of all managers throughout the organization. Recognition of this problem has led to the development of the concepts of strategic management and their incorporation in formal planning procedures. A conscious effort is needed to define a corporate identity which manifests itself in the formulation of strategies, objectives and policies. But these strategies will evolve with time to take account of new opportunities or threats often a consequence of technological change. All too often the importance of R & D in this process is insufficiently recognized.

Why should this be so? There are a number of reasons for this relative isolation of R & D but the uncertainties inseparable from this activity, the long timescales involved and the challenge of creative ideas, all mark R & D as different in some important aspects from the rest of the company. Furthermore, much of the work of R & D is concerned with future corporate developments frequently involving changes to the products, processes and management practices upon which past success has been built. Change represents a threat to the established orthodoxy and is likely to arouse suspicion and opposition from many of those affected by it. This can lead to the isolation of R & D from the mainstream of corporate activities.

The technologist must recognize that the fault is not always on one side. Whilst giving lip-service to commercial needs, he may derive personal satisfaction more from his continuing contribution to a project than the achievement of its business objectives. Like Stevenson, he finds 'it better to travel hopefully than to arrive'. In contrast, the businessman tends to be a man of action rather than a thinker.

He may have little understanding of technology, or sympathy with the problems of the technologist, although he respects his professional ability.

Thus we have a managerial gap which it is essential to bridge if technology is to make an effective contribution to the business. This is a gap of understanding, attitudes and motivations. To close it, the businessman must know more about the potential of technology and the process of innovation, and the technologist should influence and be influenced by corporate considerations.

With the advent of the new technologies, particularly microelectronics and biotechnology, the pace of change is accelerating both in society and in the individual organization. The full impact of these technologies is often not easily understood by the technologist himself. It is even more difficult for the average member of the public and the non-technological manager. Thus there is resistance to the new technologies, partly based upon ignorance and partly as a result of genuine concern for possible secondary effects difficult to assess. The R & D manager considering the influence of technology upon his business is, therefore, concerned at three levels. First, he must look outside his organization and his own area of specialist competence to identify future potential with due regard to those social, economic and political factors which will influence the pace of development. Secondly, he must ensure that the contribution of R & D is fully recognized within his organization as a whole and that the projects he undertakes are related to and integrated with the plans for corporate growth. Thirdly, he must manage the resources allocated to him to provide new products and processes in a cost-effective and timely fashion.

This book addresses these problems. It is not intended to be a handbook of R & D administration. Nor is it concerned with detailed descriptions of management techniques. It is directed towards broader issues. It focuses on those areas of management where the technology of a company relates to the firm as a whole. Thus it is written for senior managers in all areas of the business, be they in top management, marketing, finance, production or R & D. In particular, it is hoped the readership will include corporate planners who must rely increasingly upon the contribution of technology for the accomplishment of their objectives. Of growing importance too, are the business graduates for whom the management of technology is now being introduced into their curricula. Earlier editions of this book have been widely used in Europe, the USA and elsewhere as a course text on postgraduate MBA programmes and short specialist courses for senior R & D managers, and corporate planners.

This fourth edition has been amended to incorporate the findings of the growing body of research into the subject. Valuable contributions have also been made by those with whom I have discussed my ideas, both in the industrial and academic worlds. A debt is also owed to those numerous managers in many countries who have participated in spirited classroom discussions – to those many friends I am indeed grateful.

B.C.T.

Ilkley, August, 1991

Acknowledgments

We are grateful to the following for permission to reproduce copyright material:
Basil Blackwell & Mott Ltd for table on p. 36 of *R & D Management*, Vol. 1,
No. 1 (Oct. 1970) by Bell and Read, figure on p. 24 of *R & D Management*,
Vol. 1, No. 1 (Oct. 1970) by D. G. S. Davis, an extract from *R & D
Management*, Vol. 1, No. 1 (Oct. 1970) by T. J. Allen; The Institution of
Chemical Engineers and Authors for an extract from Conf. Proc. 'Productivity
in Research' (1963) by A. Hart and an extract from 'Evaluating Signals by
Technological Change' by J. R. Bright in *Harvard Business Review*, Jan./Feb.
1970; Financial Times Ltd for an extract from an article by R. Whitaker in *The
Financial Times*, 10 February 1970; Microforms International Marketing
Corporation for quotation from *Long-Range Planning*, Vol. 1, No. 4 (June
1969) by K. G. Binning and a quotation from *Operational Research Quarterly*,
Vol. 20, No. 3 (1969) by D. J. Williams; Management Centre, University of
Bradford for extracts from Conf. Proc., 'Management of Technological
Innovation' (1969) by J. R. Bright and a figure on p. 202 by H. Dewhurst in
Research Management, Vol. III, No. 3 (May 1970); The Royal Society and
author for an extract from *Proc. Roy. Soc.*, 1969, A.314 by L. A. B. Pilkington.

We are also grateful to the following for permission to reproduce illustrations:
Carfax Publishing Company for Fig. 2.12 from 'Globalization of Technological
Innovation' by R. Petrella in *Technology Analysis and Strategic Management*,
Vol. 1, No. 4, 1989; Pergamon Press Ltd for Fig. 2.13 from Fig. 3 in article
'Co-ordinating Business Strategy and Technical Planning' by A. C. Frohman
and D. Bitendo, p. 60 of *Long-Range Planning*, Vol. 14, No. 6, December 1981;
Fig. 6.4 from 'Research Planning Diagrams' by D. G. S. Davies, *R & D
Management*, Oct. 1970; Fig. 8.9 from *Technological Forecasting for Industry
and Government; Methods and Applications*, © by James R. Bright, Prentice-
Hall Inc.

Introduction

Technology and society

Technology is the major stimulus for change in society. The world we live in has been fashioned by man's application of the knowledge acquired over the centuries. It is this which distinguishes us from other animals. Yet it is only over the past two centuries that the application of technology within industry has become a major force in society. Research and Development as we know it today, absorbing a sizable proportion of corporate funds, is much more recent and is mainly a post-Second World War phenomenon. It represents the purposeful and systematic use of scientific knowledge to improve Man's lot even though some of its manifestations do not meet with universal approval.

For about thirty years after the war economic growth founded largely upon the development of technology-based consumer products was rapid. However, since the mid-1970s there have been a number of shocks to the world's economic system. The immediate cause of these disruptions was the series of oil crises which were themselves a consequence of the rising demand for oil-based products and transport so essential a part of the technological revolution of the post-war period. At the same time we have witnessed the emergence of a new range of technologies based upon microelectronics and information technology heralded as the dawn of the post-industrial society. This term is, however, misleading since it relates to human activities rather than the role of industry in providing the products and services which will still be demanded in ever increasing quantities. Whatever the future form of society the contribution of technology and R & D can be expected to increase. There seems little doubt, however, that we are now living through one of those infrequent periods in history when human society undergoes a profound change in its direction.

It is not the purpose of this book to prognosticate about the future, but the uncertainties inherent in so much of what we read and talk about these days is of central importance to any consideration of technological innovation. If we are no longer sure of what the future will hold, it becomes increasingly difficult to manage any activity orientated towards the future. For we are now concerned with two dimensions of uncertainty − that of the innovation itself, and of the environment into which it will be launched at some future date.

It is frequently stated that change today is more rapid than in the past. But change itself is not a new phenomenon. Technologists must appreciate this and realise that the period of steady growth to which they became accustomed is not typical of history. They must accept this as the reality and ensure that their efforts are directed in a way which is relevant to the emerging needs of the future. In this context they will have to focus their energies towards applying the new technologies to new ends. Most of the growth industries of the past are now approaching maturity and offer limited scope for further expansion, particularly in the developed world where the markets are approaching saturation for a wide range of products. But it is in these industries where most R & D technologists are employed. If their organizations are to survive and prosper they must find new opportunities for the exercise of their innovative skills. For many it will provide a challenge difficult to meet. Another consequence is that their contribution to corporate success will extend beyond their parochial interests within R & D. Frequently only they can interpret the significance of the new technologies and the threats or opportunities arising from them. This strategic role of R & D is of increasing importance but insufficiently recognized in many organizations.

It is instructive to consider what is now happening in a historical perspective. The first use of tools, fire and the wheel had as profound an influence on the human condition as any modern advances. We shall never know the thought processes leading to the early discoveries or how long it took before they were widely adopted. Later the invention of the printing press laid the foundation for the widespread dissemination of knowledge leading to universal education and the breakdown of the feudal system. However, industrial society as we know it today really started with the Industrial Revolution which was based upon the harnessing of power to transform manufacturing processes. The initial impact was primarily in relation to production operations rather than the development of new products, which remained relatively unchanged. Only in the last century through electricity, industrial chemicals and the internal combustion engine did the focus shift to the development of entirely new products, many of which replaced activities previously performed by the individual. In many of these applications of technology the scope for developing radically new products is now severely limited. Of course, demand for existing products and their derivatives will remain and there is still room for enhancing their performance. In general, however, the demand in the advanced markets is largely for replacements, many of which last longer than those they replace thereby decreasing the size of the total market. But as this phase in history draws to a close the potential for microelectronics has yet to be fully exploited. Beyond this we have the exciting prospects for biotechnology and new materials.

Thus we see a changing pattern of technological evolution. In recent years the work of Kondratieff, who suggested the existence of a long wave of economic development lasting about fifty years, has been gaining increased recognition. This has been stimulated by the depression—recession of the 1980s following half a century after that of the 1930s. Mensch,* who investigated the frequency

* Mensch, G. *Stalemate in Technology*, Ballinger, 1979.

of technological innovation, also identified a fifty-year cycle with bursts of innovation occurring at about the same time as the depths of the depressions. On each occasion different technologies have been responsible for the formation of the new industries on which renewed growth has been based. Throughout these cycles the importance of technological innovation as the stimulant of economic prosperity has been growing. There is little doubt that this will be equally true in the future. But it must be harnessed effectively. To do this it cannot be divorced from the complex economic, social and political system within which it operates.

Technological innovation today

The history of research into technological innovation stretches back to the 1950s. This brought a recognition, at least amongst academics, that it was an activity which must be managed. But before there can be effective management there must be an understanding of the process of innovation and its problems, the development of a conceptual framework, and guidelines for action. Many of these pioneering studies are as relevant today as they were then. Some of the problems which exercised the minds of these researchers, such as project evaluation methods, are no nearer a solution now than they were at that time; perhaps the uncertainties inseparable from innovation imply that there are inherent difficulties for which a satisfactory solution will never be found. The lessons from other studies, particularly in relation to organization and technical change, have not been learnt; they are not applied widely and each generation of R & D managers attempts to tackle the problems *ab initio*. This suggests a lack of professionalism amongst technologists when they move into management. The majority are not trained to manage innovation, nor do they read widely the large number of journals which have proliferated in recent years.

In the late 1980s the need to train all managers, not only technologists, in the management of technology became more widely recognized, perhaps stimulated by the threat posed by the success of Japanese industry. In this context, however, it is important to draw a clear distinction between the management of technical activities and the management of technology, the knowledge resource, as a means of generating corporate wealth and competitive advantage. This important distinction is not sufficiently understood. The training and development needs are quite different.

If the management challenge in the past was demanding, it is even more so today. Let us briefly review some of the developments of recent years which have added complexity to what was always a difficult task.

Internationalism

All companies are now exposed to developments throughout the world. Knowledge is generated in universities, research organizations and company laboratories in an ever increasing number of institutions and nations. Thus the catchment area for new knowledge has widened and a technological threat may emerge from quarters not previously considered. It is a management task to ensure that access to these developments is gained rapidly. Thus one sees the establishment of company laboratories in several countries; this too must be managed.

Not only are major firms finding it essential to become multinational, they are also realizing that it is necessary to form joint ventures, engage in collaborative developments and make acquisitions crossing national boundaries. Frequently the formation of these relationships has been stimulated by the need to acquire technological capabilities beyond the expertise of each of the partners.

Competition in the market-place is also becoming more intense and international. Many companies have suffered from imported products — often from Japan and the other rapidly developing economies of the Far East. Both their products and manufacturing technologies must be matched. In this climate only the best and most professionally managed companies can hope to survive. No longer can the second rate hide behind geographical or protective barriers.

Technological diversity

At one time most companies were primarily concerned with only one technology. This is no longer true. All companies, for example, now need a microelectronic competence as well as their traditional technology. But innovations are increasingly dependent upon the interactions between technologies, sometimes complex as in opto-electronics, mechatronics and other hybrid technologies. But as the body of knowledge in a technology grows the training of the technologist becomes more specialized. This makes it increasingly difficult to acquire the wider multi-disciplinary knowledge essential if the potential interactions are to be identified and exploited. Nevertheless this is a challenge which cannot be evaded. This suggests there is need for a 'general technologist' analogous with the general manager who has sufficient knowledge for taking decisions without having the detailed understanding of the specialist.

The need for urgency

With the increase in competition it is essential to reduce the lead time between the generation of new technology and its incorporation in new products. This gives rise to the need for 'just-in time technology' demanding a much closer integration of R & D with production and marketing than has often been customary. Not only must the technologist complete projects quickly in a form acceptable to the market and easy to produce, but he must get it right first time since shortening product lives give little opportunity for rectifying early shortcomings.

In order to respond rapidly there must be an effective process of technology capture whereby new knowledge is identified and introduced into the company in a timely fashion. This too must be managed. Nor can the incorporation of this knowledge in ideas for new products be left to chance. Efforts must be made to foster creativity within the organization.

Market diversity

Relationships with the market are becoming more complex. The identification of one technology and its products with one market is breaking down. Thus biotechnology has applications in the food, agricultural and pharmaceutical industries whereas the company with the technological competence may have experience in only one of these markets. Thus it may become necessary to forge links with other companies, sometimes through joint ventures, in order to exploit the synergy where one can provide the technology and its partner the market expertise.

Customers are becoming more selective as they become more affluent. With increasing segmentation and the evolution of niche markets, often based on adding value through technology, there is need to focus more precisely on the consumer's needs and to offer a wider diversity of products. As computer integrated manufacture (CIM) provides the flexibility to meet these market demands, so does it impose new requirements for responsiveness, flexibility and diversity in R & D and product design.

The range of consumer desires is also widening. Environmental concern, for example, raises many issues and opportunities where technology can contribute. This could include the design of 'green' products, those that can be recycled, the elimination of pollutants and the disposal of toxic waste.

The Japanese syndrome

The Japanese have become dominant in a number of technological industries. In the 1980s this was largely due to their manufacturing technology and expertise; today and in the future it will become increasingly based upon the development of innovative products such as the camcorder. This success is founded on long-term thinking supported by technology forecasting, expanding R & D funding, an obsession with growth rather than short-term profit, and a high investment in technical and managerial training. The author's research in Japan indicated that they are doing nothing particularly clever. What distinguishes them is their professional approach to the management of technological innovation.

This brief description of some of the most important trends in the corporate environment of technology indicates the complexities of the challenges facing those with the task of managing this activity. These and other factors are leading to a less structured approach to R & D within the firm. Many of the traditional barriers are falling. It can also be seen that they raise strategic issues that cannot be resolved entirely within R & D. This supports the view that in the successful company of the future we are likely to see the technologist much more closely integrated in the totality of the business. But this role cannot be discharged adequately unless he has the commitment and training.

Technological innovation extends well beyond the management of technical activities. Technology is a vital corporate resource. It has major strategic implications which demand a careful consideration of how it can contribute to the achievement of the aims of the business. Senior technologists must play an active part in the formulation of these strategies. However, these objectives can only be attained by identifying projects which combine the most appropriate and latest technology with clearly defined market opportunities. Finally, these projects must be brought to a successful conclusion. This goes beyond the completion of the project to specification and to time, for the innovation process is not complete until the product has been manufactured and marketed, something which cannot be achieved without a close organizational integration of R & D, design, production and marketing. All this demands a high degree of both technological and managerial professionalism.

Technology and strategy

In the past adaptation to changing needs has to a large extent been brought about by the decline of the older industries and the formation of new businesses to exploit the new technologies. This is widely recognized and governments actively encourage the entrepreneurs who establish their own companies. Desirable though this may be, it is only part of the answer. Today's large companies are the main generators of national wealth. Only they can mobilize the financial resources needed to exploit new developments in many technologies. They must learn to adapt to these changing needs. Many are doing so, but there are also many former industrial leaders which have failed to survive notwithstanding heavy R & D spending. All too often this is because the resources invested in R & D have not been focused to the long-term needs of the organization.

Strategy is concerned with providing a sense of direction and coherence to corporate policies and actions. It manifests itself in the allocation of resources between conflicting claims. Most importantly it should indicate where change is required in order to further the broad aims of the organization into the future.

The formulation of a corporate strategy is an intellectual process whereby the forces shaping the environment of the company are analysed and matched with its ability to meet the threats or exploit the opportunities identified. Because it is future orientated a view of the future must be taken; forecasting, including technology forecasting, has an important role to play in assisting in the development of that view. It is evident that technology is one of the most important forces shaping the future – for many companies it is the most important. Often this is not sufficiently recognized. In such cases the corporate strategy is developed by top management with inadequate technical representation. Thus it is determined 'on high' and is presented as given to the R & D department to implement its part. This can arise because of a failure of top management to appreciate the importance of technology or because the senior technologists involved accept a specialist and subservient role isolated from the wider interests of the business. Technology must be regarded as an integral part of the business, not as a tool to be used to meet predetermined corporate objectives.

There is also need for an R & D strategy. This leads to a determination of how the R & D resources should be deployed. There are never sufficient funds to do all that is desirable. Choices must be made. Should the emphasis be on new or current technology? on new products or the incremental improvement of existing products? on products or processes? and so on. These are not easy decisions to take. They must be made within the structure of the corporate strategy but bearing in mind the long-term needs of technology itself. The linkage between the corporate and R & D strategies must, therefore, be a continuous and iterative process, for the latter may well reveal opportunities which might warrant a modification of the former.

Whilst strategic thinking is important it can become an arid intellectual exercise if it is used solely to justify the status quo or is not translated into effective implementation. It can only provide a framework within which R & D must exercise its creativity and development skills. These operational tasks will form the major part of the technologist's working life.

The management of R & D

It must be recognized that R & D expenditure is an investment for the future of the organization, but that it has to be justified in terms acceptable to top management. The absence of rigorous appraisal in the past has often led to costly failures. Yet it must also be said that a critical evaluation demanding a detailed justification can rarely be provided to satisfy managements unwilling to accept risk. The uncertainties inherent in R & D are such that many of the economic justifications are to a great extent meaningless. It is a truism that if all the details of a proposal were known at the outset it would not in fact be an R & D project. Thus we must reconcile the need for faith and a good measure of optimism about the eventual success with the acceptance of the necessity for corporate control. This is a dilemma to which there are no easy answers and poses the most difficult problem facing the R & D manager.

Uncertainty means risk. Thus the attitude of top management to risk provides the framework within which all R & D decisions have to be taken. But although risk cannot be avoided it can be assessed. Risk analysis and technology forecasting techniques assist the R & D manager in evaluating the characteristics of his programme and its possible impact on corporate well-being. One must aim, therefore, to construct a balanced portfolio of projects wherever possible. In the sequential evaluation of individual projects the importance of their contribution to the total programme should not be lost sight of.

The management of an R & D programme is a complex process. There are no easy answers. Furthermore, the R & D manager often receives little preparation for the task he faces. Technical competence, although remaining important, is only one of the many attributes contributing to success. In recent years an increasing volume of research projects into technological innovation has led to a better understanding of what is involved. They do not, however, provide simple prescriptions of universal application. For convenience the factors contributing to success may be considered under the following headings:

1. *A strategic approach.* This we have already discussed.
2. *The human factor.* The R & D manager must develop effective teams within which each individual is able to make his maximum contribution. This includes such aspects as creativity, motivation and communications. The design of an appropriate organizational structure is also important, particularly in so far as it promotes effective integration of R & D with other corporate functions.
3. *The use of professional management techniques.* Many techniques both quantitative and subjective are an essential part of any management development. They include methods of project evaluation, control techniques, risk analysis, financial assessment, budgetary control and so on. These are important and should be applied wherever possible. Yet they must not be regarded as an end in themselves.
4. *A future orientation.* Timing is of the essence of innovation, which must produce the right product or process at the right time. The capture of technological advances and their assessment with the aid of technology forecasting is an important element in this.

The many case histories of success and failure in innovation provide valuable insights into what happens in practice. Supported by research into specific aspects of the innovation process they help the manager to avoid those errors which almost inevitably lead to failure. Although they reveal factors necessary for success they are less helpful in aiding his decisions in his unique environment. Unfortunately for the theorists there are too many cases of failure where there should have been success and vice versa. This may be because the theories are not yet sufficiently precise. More likely, however, is the importance of the dedication and drive of the managers responsible for the projects.

Thus researchers in this field have been properly concerned not to make definitive statements concerning a process which is imperfectly understood and which they are unable to prove with the rigour normally expected from research reports. Managers, however, are constantly faced with situations where their understanding of the underlying processes is incomplete. In spite of this they cannot avoid the necessity of taking decisions. In writing this book, the author has attempted to derive working hypotheses based upon the research findings currently available. Frequent reference is made to this research in the text and the references at the end of each chapter are carefully selected to guide the further study of the reader. The aim has been to help those in managerial positions faced with taking decisions. The reader must, however, remain critical.

Concepts, procedures and techniques all play their part in successful innovation. But they are of little value in the absence of managerial judgment and individual commitment. Nothing written in this book is intended to detract from this essential ingredient of success. The author will be well satisfied if he has stimulated the reader to a better understanding of technological innovation, the development of his own thinking and concepts appropriate to his individual business problems.

1

The process of technological innovation

A unique chronological process involving science, technology, economics, entrepreneurship, and management is the medium that translates scientific knowledge into the physical realities that are changing society. This process of technological innovation is the heart of the basic understanding which the competent manager, the effective technologist, the sound government official, and the educated member of society should have in the world of tomorrow.

James Bright

The need for a conceptual approach

Technology has been and will remain the prime stimulus for change in our society. Our major industrial companies owe their origin and their continued existence to the successful application of technology in evolving new products and improved manufacturing processes. Nowadays, when it is fashionable to attack technology because of its effect upon our environment, we must not undervalue its continuing contribution to the quality of twentieth-century life. While the environmental effects should not be disregarded they must not be accepted as a condemnation of technology. They do, however, show clearly the challenges which have to be faced in the future. Thus the ends to which we put our technology may change, but its importance will not diminish. Above all the need to ensure that it is effectively managed in meeting the needs of society will become more pressing as it absorbs an increasing proportion of our national resources.

For many organizations, built up on past success, the impact of technology can present a threat. Companies which have failed to maintain their innovative momentum have been overtaken by more youthful and vigorous organizations. Not infrequently they have gone out of business. A comparison of today's industrial leaders with those of twenty or even ten years ago shows how many of the once great names have declined in importance or disappeared from the business scene. In many cases this was caused by their inability to anticipate the effect of a new technology whereas their competitors had seized the opportunity for growth which it had offered.

Our task would be easier if we could conclude that business death followed solely from the absence of innovation. Unfortunately this is not so. An analysis of business failures reveals amongst them a significant number of innovators who have failed to translate their technological creativity into profitable business operations. The challenge, then, is not only one of innovation but of *managing technological innovation for profit.*

Much effort has been devoted to attempts to justify R & D expenditure. This is not easy since there is no satisfactory method for measuring R & D output. Many researchers have used the number of patents since this is quantifiable. But the validity of this measure is questionable due to the near impossibility of relating this output of knowledge with its commercial application, the difficulty of assessing the 'quality' of patents as distinct from their number, and the different patenting policies of companies in the same industry. Ideally one would like to correlate R & D expenditure with profitability. The evidence that this can be done is not conclusive. More convincing is its positive relationship to company growth [1]. This raises an important issue for management. R & D investment which is for the long term inevitably reduces short-term profitability. There is increasing concern that this 'short-termism' in several Western economies has been a major factor in their relative decline compared with Japan where growth based upon a high and increasing funding of R & D is a prime corporate objective [2].

It would be folly to pretend that decision rules or techniques can be developed which will ensure success in an activity where uncertainty is so much greater than in other areas of business. Managers cannot be taught how to manage technology successfully. But this does not mean that it is impossible to develop their ability to make better decisions by exposure to the theoretical techniques and concepts of management thinking.

There is a clear distinction between science and technology. Few industrial organizations have the need or the resources to carry out basic or fundamental research aimed at acquiring new knowledge for its own sake. Although the words 'science' and 'research' are often used within industry their use is more an inheritance of university training than a true description of the work undertaken in industrial laboratories. Price [3] stresses the importance of scientific research as the provider of the knowledge that business exploits, particularly in high technology industries. The industrial technologist must capture this scientific knowledge generated elsewhere; it is not his job to do the research himself whatever his personal inclination.

It is also useful to draw a distinction between the terms 'technological innovation' and 'research and development' although it is commonly supposed that they cover the same range of activities within a company. But technological innovation implies a company-wide approach to the profitable application of technology rather than a description of the activities of one department responsible for R & D; it stresses the importance of the whole innovation process through to commercial exploitation; and it leaves the door open to new technology which originates outside the company. For the only justification for devoting scarce financial resources to research and development is the belief that they will generate innovations which will contribute to the company's survival and continued profitability. Furthermore, it must lead to the attainment of these objectives more cheaply than if the money were spent in some other way. Convincing answers must be given to such questions as:

1. Can the objective be achieved more economically by licensing another organization's technology than by initiating an internal R & D project?
2. Will investment in R & D to develop new products and processes yield a greater return than investment in manufacturing or marketing?

These cannot be answered easily since they give rise to broader considerations such as:

1. Can the potential of licence agreements be recognized without a strong in-house R & D capability?
2. While investment in manufacturing and marketing may yield substantial short-term benefits, can they ensure survival in the long term?
3. How accurately can past R & D investment be correlated with commercial success now?

In spite of the difficulty in answering these questions conclusively, the technologist must not lose sight of such considerations, for the ultimate financial outcome is the only justification for his employment in industry.

Discussion of the problem is further complicated by the wide variety of forms which technological innovation can take. A radical technological innovation is likely to be associated with a much higher degree of uncertainty than a minor product improvement, although its potential rewards may be correspondingly greater; a product innovation has a different impact on the company's operations from a process innovation; the characteristics of innovation in the chemical or pharmaceutical industry are not the same as in aerospace, electronics or engineering. While these distinctions should not be disregarded, they tend to be differences of degree rather than of character. Most of the examples which we shall examine relate to radical innovations in new products because it is with these that the difficulties are most acute. However, the concepts which apply to these situations of maximum uncertainty are still appropriate where the issues are more clear cut. It is in the managerial techniques which can be employed to implement the concepts in these varying situations where the major differences are found.

Thus the thesis upon which this book is based is as follows:

1. Technological innovation is a critical factor for the survival and growth of most industrial enterprises, and should not be left to chance if it can be planned and controlled in a meaningful way.
2. Resources devoted to technological innovation can only be justified in so

far as they further the attainment of corporate objectives.

3. Analysis of past technological innovations reveals a number of factors all of which appear to be present in many successes, and one or more of which are found to be frequently absent in failures.
4. A conceptual approach and an understanding of the processes at work can be developed.
5. Improved decision-making and higher return on the investment in technology should follow from a conscious attempt to apply the concepts in practice.

It must be stressed that there is no suggestion that the application of the concepts discussed in this book will automatically lead to success. The process of innovation is far too delicate for this ever to be the case. But we are considering an activity in which failure is the norm and where the greatest part of the investment in technological innovation is wasted. Wind and Mahagan [4] conclude that in spite of the increased professionalism of the past 20 years: 'There are relatively few truly innovative new products and, most disturbingly, the success rate of new product entry is still abysmal.' Commercial success rates are frequently as low as 10% of the projects initiated − put the other way this means a 90% failure rate. There is ample evidence to show that many of those failures were avoidable. Thus a modest reduction in the failure rate can yield a significant improvement in the productivity of resources devoted to R & D.

Technological innovation as a conversion process

Managers devote most of their time to dealing with tangible situations. Concern with such problems as the marketing of an established product, the manufacture of a product to a specified design with a given range of tools, or the progress of a clearly defined R & D project, leads to a preoccupation with 'things' rather than 'ideas' or 'concepts'. Industry is perceived as a process which converts raw materials into products, which in turn are converted into money by sale to customers who are prepared to pay a higher price for these goods than the cost of producing them. The margin between the price paid and the cost to produce yields the profit which can be reinvested in the company to sustain its growth or returned to shareholders in the form of dividends.

The conversion process illustrated in Fig. 1.1 shows the predominant product orientation which was widespread at a time when the customer's position in relation to the manufacturer was weak. As the balance of power has shifted towards the customer so does the need to pay greater attention to his

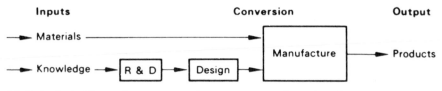

1.1 Technological innovation as a conversion process − product orientation

requirements. The 'marketing concept', born out of this change of emphasis, propounded that products were not an end in themselves but were merely the means of satisfying the customers' needs or desires. 'Market research' grew in importance and products, particularly consumer and consumer durable goods, came to reflect the requirements of the customer more closely. But this process, although a step in the right direction, failed to bring the initiator of the product, the technologist in the R & D department, into direct contact with the user.

Thus the technologist working in his laboratory remained largely isolated from market forces. This was often particularly true in the case of radical new technologies where the intellectual excitement of the technology itself could blind the technologist to its limited commercial potential for satisfying an identifiable human need.

An alternative approach is shown in Fig. 1.2. This considers the process as being the conversion or transfer* of scientific or technological knowledge directly into the satisfaction of a customer need; the product then becomes merely the carrier of the technology and the form it takes is only defined after the technology and the need have been clearly matched.

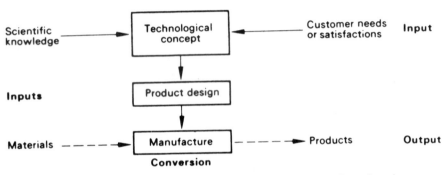

1.2 Technological innovation as a conversion process – technology/market orientation

At first sight the difference between these two approaches may appear to be academic – it is, after all, the product for which the customer finally parts with his money. Yet it represents a significant change in the technologist's attitude towards the conversion of his scientific knowledge into a business purpose, for it focuses his mind on the needs of the potential customer. This requires a greater identity of thought between himself and the marketing manager than is customary in many companies and may involve organizational changes to make it possible. Furthermore in the case of a radical innovation he may find it necessary to bypass the marketing department because its traditional expertise and knowledge which relate to existing products, markets and

* In recent years the term 'technology transfer' has been used widely. There is, however, no universally accepted definition. Some authors restrict its use to the transfer of technology between developed and less developed nations. Others widen the use to cover all aspects of the transfer of technological knowledge into society. In view of this ambiguity the words 'technology transfer' will not be used in this book.

customers may be of little assistance when the potential of the new technology can only be realized through radical new products and markets. In these circumstances the conventional wisdom and limited horizons of the marketing manager may well lead to the wrong conclusions.

Premature thinking in product terms must be avoided since it can restrict the search for applications of a new technology to those which are most immediately apparent. These frequently are not those with the greatest commercial potential. This conclusion is supported by research into thirty technological innovations by Bright [5] who states:

> The most important application of a new technology is not always that which was visualized first; and as a corollary, technological innovations frequently gain their first foothold for purposes that were originally not thought of or are deemed to be quite secondary.

and:

> The sponsors of a radical technology should adopt a policy of searching for applications with an open mind towards new uses and a readiness to support trial in unexpected fields. The strategy should be one of exploration, rather than of single-minded commitment to one predetermined usage.

Bright attributes a great deal of this uncertainty to the difficulty of forecasting the technical and social environments into which the new technology will be launched some years later. This difficulty must be recognized, yet there is little doubt that the very high level of failure stems at least in part from the absence of a systematic attempt to match the attributes of a new technology to future market needs.

Value added by technology

The satisfaction of the functional needs of the market, although a prerequisite for success, is only one of several ways in which technology can contribute both to the customer and the business.

The customer is concerned with all aspects of the product throughout its life. These include durability, reliability, in-service life, maintainability and service in all its aspects. For example a new drug delivery system which enables smaller quantities to be targeted on the seat of an infection may be as important as the properties of the drug. Improved packaging or containerization can also provide a competitive advantage. Innovation in all these areas and systems, often by exploiting technological potential, to make the product more user-friendly must be considered an integral part of the technological innovation process.

Profit, a central corporate concern, is dependent on the price that can be charged and the costs of production. Cost can be reduced by the use of new materials, a reduction in the quantity of material used, design to minimize the number of components and their ease of manufacture and assembly, improved production processes and more effective managerial systems. Thus the exploitation of technology extends beyond the performance characteristics of the product itself. The technologist must at all times seek ways in which he can apply his technological knowledge to further both the internal needs of the company as well as those of the customers of the business. An obsession solely with the product itself, a natural tendency, must be avoided.

Although attempts should always be made to eliminate unnecessary costs, cost reduction is often not of overriding importance in gaining competitive advantage. In affluent societies high value added products designed for niche markets usually command high profit margins. It is technology, for example in four-wheel steering or ABS brake systems, which adds value.

In recent years a great deal of attention has been focused on manufacturing technology. CAD/CAM, for example, has improved responsiveness by reducing the time from product concept to manufacture. However, this is only one part of the innovative chain extending from the acquisition of technological knowledge to its delivery to the customer incorporated in the product. Far less attention has been paid to reducing the time throughout the innovation chain. Just-in-time concepts are as valid for knowledge as they are for materials. The value is only added when it is delivered to the customer.

Factors contributing to successful technological innovation

Research into the causes of success and failure in technological innovation reveals the frequent presence of a number of factors. One is tempted to write that in the absence of one or more of these factors an innovation is doomed to failure. Although this undoubtedly overstates the case, there is now a substantial amount of evidence to support their importance. There will always be exceptional cases where the benefits accruing from an innovation are so outstanding that it succeeds in spite of poor management. But these are exceptions and their occurrence does not absolve management from attempting to ensure the presence of all those factors which have been shown to be important in a wide range of successful innovations.

The most critical of these factors are:

1. A market orientation.
2. Relevance to the organization's corporate objectives.
3. An effective project selection and evaluation system.
4. Effective project management and control.
5. A source of creative ideas.
6. An organization receptive to innovation.
7. Commitment by one or a few individuals.
8. A production orientation.

Market orientation

In discussing technology as a conversion process the importance of establishing market needs has been highlighted. A useful distinction can be drawn between the words 'invention' and 'innovation'; the difference has been defined as:

INVENTION   — **To conceive** the idea
INNOVATION — **To use** — the process by which an invention or idea is translated into the economy [6].

Thus for an 'invention' to become an 'innovation' it must succeed in the market-place. Although this may appear a truism, the linking of R & D and marketing in a meaningful partnership poses one of the most serious practical barriers to successful technological innovation even when the need for it is appreciated at an intellectual level.

This statement needs amplification for there is little doubt that many industrial technologists would strongly deny that this is still true today although it might have been valid in the past. Yet in spite of the considerable improvement in recent years there is substantial evidence from both sides of the Atlantic that a market orientation is still woefully absent in many decisions and that this is a major source of failure even in the most technologically advanced companies. This is not intended as a condemnation of the technologist, for the difficulties to be overcome are very real and stem from the education and value systems of both technologists and marketing managers as well as the environments within which they work. Two major problem areas can be identified:

1. Communication difficulties between technologists and marketing managers.
2. Company organizations which hinder rather than assist effective communication between them.

The continued importance of these communication barriers is supported by the replies given to the question, 'What do you consider to be the major barriers to innovation within your company?' which formed part of a questionnaire completed by 175 research directors and senior research managers attending courses at the University of Bradford. Of these 72% referred to communication with or their relationship with marketing and marketing managers. No other factor was mentioned by more than 20% of the respondents. This sample was largely drawn from members of large research organizations in technologically intensive industries. In smaller organizations, where the coupling between departments is likely to be closer, these difficulties may not be so serious.

Similar informal studies with senior marketing managers attending courses at the same university indicated that their prime orientation was outside the company towards the customer. Their role as the main channel for providing regular market information to those responsible for the development of new products was regarded as being of secondary importance.

Ideas for new products normally originate in one of two ways — from a specified market need or from the R & D department. Within any company there is frequently a wide difference of opinion about which of these sources is the more fruitful depending upon the functional orientation of the person stating his views. However, research projects into the stimuli for innovation indicate that between a quarter and a third of the ideas originate within R & D but that their importance may be greater than this proportion suggests since these tend to be the radical or major innovations which open up new fields and eventually lead to additional need-orientated programmes. Whatever the exact proportion, it is clear that both R & D and marketing are major sources of ideas the development of which can be inhibited by poor communications and lack of understanding. In recent years the term 'technology push' has often been used in a critical sense as distinct from 'market pull' which is widely advocated. Many Japanese companies, however, actively pursue a policy of technology push. Marsh [7] quotes the examples of Canon, the camera manufacturer, whose 'view of the transfer process starts with the need to master key areas of technology. Useful products will then flow to a large extent automatically,' and of Honda, 'which does not have a marketing department; most marketing functions such as the analysis of future customer requirements are left to the company's research staff.' This approach is not likely to be acceptable to most Western companies where the need to establish a firm linkage with the market is regarded as a prerequisite for an investment in technology. Nevertheless it does indicate that both technology push and market pull have an important part to play in successful innovation. A meaningful linkage between technology and the market is required; this can only be achieved by a close relationship between the people who have the expertise in the two fields, the technologist and the marketer. It is, for example, not uncommon for R & D managers to appreciate the potential market applications for their technology but to fail to convince the marketing department who may reject it on insufficient evidence or because it does not fit their short-term product-market policies. Similarly technologists may reject product ideas originating within the marketing function on the grounds of poorly substantiated technical objectives.

Differences in personal values partly explain why communication between research managers and other managers in an organization tends to be less effective than between any other two groups of managers. The technologist regards himself primarily as a professional rather than a businessman. His training and natural inclinations enable him to relate with similar professionals in other companies more easily than with other managers in his own organization. Sociological research supports this view. Lucas and Bush [8] used the Sixteen Personality Factor questionnaire (16PF) to assess the personality traits of respondents in marketing and R & D; this revealed some significant differences. Their research indicated that marketers are more dominant and self-assertive, happy-go-lucky and enthusiastic, venturesome, spontaneous and self-opinionated whereas R & D scored significantly higher on self-sufficiency. Earlier work by Margerison and Elliot [9] on a sample of 42 research managers led them to the conclusion: 'In our research amongst R & D managers, it is clear that the work itself is the prime factor of importance, but having that job in their present organization is of least importance.' This was in sharp contrast with the attitude of marketing managers who identified closely with their organization.

The entrepreneurial interests of management sometimes come into conflict with the broader field interests of research personnel. Management's concern with a strictly marketable, promotable idea may in certain instances appear superficial to the researcher who sees his own interests as being more fundamental than this. The researcher occasionally gets the notion that he is 'prostituting' himself to commercial ends. Management, on the other hand, cannot understand the researcher's inability or perhaps his unwillingness to create new ideas in the entrepreneurial sense.

These differences in personal values give rise to a culture within the R & D laboratory which may differ substantially from that found in the rest of the organization. This is not necessarily unhealthy and few would disagree that it creates the right environment for good research. But good research by itself does not necessarily result in innovation important as it may be in providing the technology upon which innovation is based. Furthermore, it can be seen that the innovator within the R & D department generally possesses personal characteristics that mark him out from the majority of his colleagues.

The persistence of these attitudes is supported by the research of Rosenau [10] into the emphasis technologists place upon four factors: exceeding performance specifications, meeting production target costs, reducing development time and keeping within the development budget. He concludes: 'Based on this sample, people working on new products place more value on exceeding the performance specifications than beating the schedule.' The situation described is summarized in Fig. 1.3.

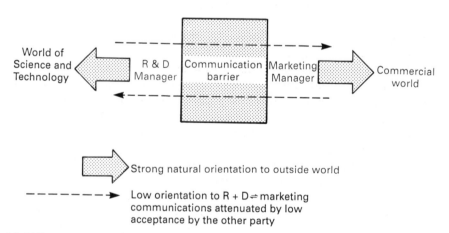

1.3 Different orientations of R & D and marketing managers create a communication barrier

Viewing innovation as a conversion process and establishing good communications between technologist and marketer are essential but not sufficient criteria for success. The product must satisfy all the requirements of the customer. If the technologist has no direct contact he may be unaware of how the product is actually used. It may not be 'user-friendly'. Furthermore, as von Hippel [11] has shown, the user himself may often modify the product to meet his needs more closely, a valuable source of innovative ideas. In consumer products the designer may assume a higher expertise than the user possesses, a frequent problem with domestic appliances. Often the designer has not operated the system himself. Hands-on experience can yield valuable insights into these practical problems. In one tractor manufacturer the author found that none of the designers had driven a tractor; this was regarded as the preserve of the trials department. Thus it was not surprising that one of its products failed in the market because it provided inadequate rearwards vision, an essential feature for farmers. Habermeir [12] has developed a framework to assist in the analysis of user product improvements.

User resistance can also provide a major barrier to an innovation, however desirable it might appear to its originator. The QWERTY keyboard was developed in the last century to slow typing speeds to avoid jamming of the mechanical components. Now that this problem no longer exists many improved layouts have been designed but have not gained acceptance due to the users' expertise and training in the existing layout. The role of the human in the total system must not be ignored.

Some companies actively discourage contact between designers and customers since this is regarded as the prerogative of the marketing department. This is highly undesirable. It must also be recognized that many technologists, particularly in large companies, are more interested in the elegance of their technical solutions than the needs of the user.

These problems can only be overcome if:

1. The designers and customers interact, with, of course, the agreement and cooperation of the marketing department.
2. Designers gain hands-on experience of using the product wherever that is feasible.

Any improvement must be achieved within an acceptance that differences in personal values and primary orientations are real. They stem from the nature of the education, skills and attitudes required to be either a good technologist or a good businessman. An attempt to go too far in turning one into the other would probably be doomed to failure, but, if successful, might possibly reduce their effectiveness in their primary roles.

How can an organization attempt to resolve this problem? Several approaches are suggested, some of which will be explored further in later chapters:

1. Development of mutual understanding through educational programmes − there is some evidence to indicate that this is already bringing some improvement in communication.
2. Organization structures which foster closer collaboration or at least avoid the creation of formal barriers to cooperation.
3. Movement of personnel − this is usually only practicable in one direction,

from R & D to marketing. This is practised by a few companies who not only find that it improves communication but also that it assists in solving possible career difficulties for the technologist with management ability who has passed his creative peak.

4. Ensuring that working with R & D personnel in new product development is specified where formal objectives are set for marketing managers.

5. Involvement of both R & D and marketing managers in the formulation of corporate and R & D strategies and in project selection and evaluation decisions.

Relevance to the organization's corporate objectives

The commercial desirability of a proposed technological innovation cannot be divorced from the organization within which is arises. Companies which may at first sight appear to be directly comparable will on closer examination reveal fundamental differences in their business and managerial characteristics. Considerable variations in business objectives, attitudes to risk and innovation, and value systems will emerge.

Rapidly changing business conditions, themselves brought about largely by technology, have made it necessary to plan for future growth to an extent which was not formerly considered essential. In the past it was possible to adapt to environmental changes when they occurred. Nowadays when change is more rapid and the gestation period for many technological products is longer it is necessary to anticipate rather than react. Acceptance of this need has led to the rapid growth of formal corporate planning.

Definition of the products and markets upon which future growth will be based is an important element of any corporate plan. In formulating a corporate strategy attention is paid to the organization's own capabilities in relation to the threats and opportunities which have been identified in the business environment. It follows that some potential new products, attractive as they may appear in their own right, are inappropriate for development by a particular company because it does not have sufficient resources or does not wish to enter or extend its operations in certain markets.

Even where there is no formal corporate plan these considerations cannot be ignored, for they will still enter into top management thinking even if not stated explicitly. No company grows in an entirely random fashion and one can always identify an implicit understanding of what its values are, what business it is in, and so on. However, in this situation it becomes more difficult for management at lower levels to relate their decision-making to the objectives of top management. Furthermore, in the absence of clearly stated objectives, strategies and plans, there is no standard of reference against which changes can be evaluated. Consequently, decisions may be taken which are no longer appropriate because they are based on assumptions which have ceased to be valid.

Corporate planning becomes an academic exercise if it operates at top management level but is not translated into decision-making at all levels of the organization. Nowhere is this more true than in R & D, for decisions taken in the laboratories today determine the future product mix upon which the accomplishment of the corporate strategy depends. The corporate plan thus becomes the standard of reference against which R & D performance needs to be measured.

Is formal planning compatible with R & D activity? To many technologists the very word 'planning' in this sense is anathema since it carries with it the implication that they are losing some of the freedom they value highly and consider essential for a creative activity. But a line must be drawn between freedom and licence for no one has the right to spend company resources on indulging his own interests. The validity of planning must, therefore, be judged by only one criterion – whether it adds to or subtracts from the effectiveness of the organization. There is little doubt that the imposition of a high degree of centralized planning on R & D can be counter-productive. But inflexible planning which stifles innovation is bad planning. On the other hand planning can make a contribution if it ensures that innovations are only accepted when they are consistent with company objectives with the proviso that attitudes are sufficiently flexible to permit modifications of the strategy in exceptional circumstances.

We have seen that the main stimuli for innovation originate as ideas from the R & D department or from the definition of market needs. In the latter case the formal definition of future market and types of product stimulates a systematic examination of potential market needs which might be satisfiable through new products. By contrast the generation of ideas within R & D is frequently much more random. Consequently, some proposals may have to be rejected, not because of any intrinsic shortcomings, but on the grounds that their development would necessitate a shift of the company's growth in an undesirable direction.

Ways in which project selection and corporate strategy can be related by means of a strategy for R & D will be discussed in Chapter 2. Here it is sufficient to stress that new project decisions are so fundamental to the company's future that they cannot be divorced from top management policies. They are too critical to be left to research management alone, for this would lead to a 'tail wagging the dog' situation in which the company may slip into costly development inappropriate to it or the waste of resources on developments which will be terminated by top management at a later date.

An effective project selection and evaluation system

Much attention has been focused on project selection and evaluation techniques. This is rightly so, for this is the most critical decision area for the R & D manager. The techniques available range from simple checklists to highly sophisticated quantitative analyses based upon operational research techniques such as linear programming.

The manager is faced with the difficult task of evaluating the wide range of techniques described in the literature, selecting a system which fits his needs, and deciding how much time and effort should be devoted to the selection and evaluation process. Research studies show that R & D managers have been reluctant to adopt many of these techniques particularly those which are highly quantitative. Why should this be so? In a later chapter we shall discuss a number of techniques and attempt to evaluate them. The main difficulty will be seen to be with the uncertain data which is fed into the evaluation system. With perfect data almost any technique would provide a sound basis for selection.

Examination of projects which fail or are prematurely terminated leads to the conclusion that many of them should never have been initiated in the first place. The reasons for failure could have been deduced from information available at the time of selection but was either ignored or not examined analytically in the context of a systematic evaluation procedure.

The conclusion to be drawn is that a formal selection procedure is essential and that R & D management in general makes too little use of the techniques available. But this does not imply that a technique can be applied as a simple routine procedure. Managerial judgment is the critical ingredient, but this judgment should lead to improved decision-making when it is applied to data systematically collated and analysed with the best tools available.

Effective project management and control

The need for effective project management is self-evident. Inadequate control results in cost escalation and time over-runs. Cost escalation may not only destroy a project's financial viability but, more seriously, in the case of a major project, it may lead to financial demands which are beyond the company's resources irrespective of the eventual economic rewards which might be expected from a successful completion of the development. Programme slippage also results in higher development costs, which may lose the competitive edge of being first in the market, and will certainly reduce the product's useful market life, since the date when it becomes obsolete is unlikely to be affected significantly by delays in initial market launch.

Management of the R & D programme on a day-to-day basis absorbs most of the working time of R & D directors and managers. The immediacy of most project management problems should not, however, be allowed to detract from considerations of longer term importance, such as project selection. If faced with a choice between evils it is better to have a badly managed project with potential than a well controlled failure.

The management of R & D presents problems which in many respects are more demanding than those found in other functions of business. The term R & D is itself misleading for it covers a spectrum of activities ranging from basic research through applied research to prototype or pilot plant development and final development of the new product and processes followed by further stages of product or process improvement. Each stage of this evolution has different characteristics and calls for a management style and the use of management techniques which may be quite inappropriate at another stage. During applied research or feasibility studies when it is usual to find only a small group of professional people working on a project under conditions of great uncertainty, an informal participative management style is usually appropriate. Similarly detailed budgetary and programme control may be unrealistic until the project has advanced to a stage when reasonable estimates of cost and time can be made. But it is the extension of this informal management style into later stages of development which so frequently leads to delay and cost escalation.

Thus a system of management is required which changes continuously as a project advances. The achievement of this is one of the greatest challenges to R & D management and an area where it most often fails, either through an inability to appreciate the need or through a lack of the necessary skills. One important factor which must be considered is whether one man can be expected to be sufficiently adaptive to cope with these changing demands or whether it necessitates changing the project manager during the course of development.

A source of creative ideas

Innovation cannot be divorced from creativity. The successful innovation is offering the market something new for which the customer is prepared to pay. This arises from a new technology or a new application of an existing technology. The quality of the innovation results from the originality or the creative minds of one or a few individuals. Without creativity there can be no innovation.

The concentration of R & D management research and thinking on selection and evaluation procedures may distract attention from the fact that these procedures can only serve a useful purpose if they receive an input of high potential proposals. It is akin to a prospector panning for gold − if there are no particles of gold in the sand, improvements to his extraction system are not going to pay any dividends.

It is commonly supposed that R & D departments enjoy a surfeit of promising project proposals − the problem is to decide which of them to support. But is this so? The evidence is mixed, but there are indications that some companies and some industries are experiencing increasing difficulty in finding a sufficient flow of worthwhile project proposals. In part this can be attributed to the variations in creativity found amongst different companies. But there is probably a more fundamental factor at work. A technology can be likened to a product with a life cycle. In its early days it opens up a whole new area of opportunity but as time progresses most of the knowledge which can be usefully employed becomes fully exploited and larger investments are required to produce marginal benefits. Using another gold mining analogy one can liken the process to the extraction of ore from the most profitable seams first. With time it becomes increasingly difficult to find new rich ore-bearing rock and the rewards from the existing workings become less profitable to extract.

Thus it is not surprising that long-established industries experience trouble in finding new innovations from their existing technology. Occasionally such an industry will be transformed by a major technological innovation; this happened, for example, in the flat glass industry with the development of the float glass process. Typically, however, an established industry has low margins of profit and is highly competitive. In the absence of major innovations progress results from a series of small advances which become increasingly difficult to obtain.

Creativity is not confined to R & D; ideas can come from anywhere and should be encouraged although many will be of little value. From within the company new product proposals can be expected from top management and marketing and ideas for process improvement from production. From outside there are three main sources − competitors, suppliers and customers. There are usually many ideas not covered by patents incorporated in competitors' products which could be adopted with benefit once the Not Invented Here barrier is overcome. Customers are a particularly fruitful source of ideas; some consumer product companies actively encourage their views (e.g. Unilever). The research of von Hippel [13, 14, 15] and others has shown the contribution of users to process innovation; in one study of semiconductor and electronic subassembly manufacture 67% of significant process innovations originated with the user. This process can be formalized to ensure the systematic approach to assessing how customers use their products, a method used by Bosch in its hand-tools division.

As more technological industries reach maturity, finding and exploiting innovative ideas will become more important. Is this a process which must be left to chance? Obviously not if there is an alternative. A later chapter will be devoted to this topic, but we can see already that if an organization recognizes this as a real problem it must take whatever steps are within its ability to create conditions conducive to creativity. The possibilities open to the company relate to:

1. The recruitment and retention of creative people.
2. The creation of a working environment which encourages creativity.
3. The use of any techniques which can be shown to be useful in developing creative problem-solving.

An organization receptive to innovation

Innovation means change. Thus it can be interpreted as a threat to people who are affected by it and is likely to arouse their opposition. The psychologist Schein writes [16]:

Organization planners or top managers often naively assume that simply announcing the need for a change and giving orders that the change should be made will produce the desired outcome. In practice, however, resistance to change is one of the most ubiquitous organizational phenomena. Whether it be an increase in production, or adaptation to a new technology, or a new method of doing the work, it is generally found that those workers and managers who are directly affected will resist the change or sabotage it if it is forced upon them.

As the major agent for change within an organization the R & D manager must expect to meet opposition to every proposal he puts forward; the more radical it is the greater the resistance he must expect. In many organizations lip-service is paid to the concept of innovaton until it begins to affect people personally. This is even true amongst managers within an R & D department − propose a change to the internal organization of the laboratory and wait for the objections. For it is always others we expect to change, not ourselves.

The objections raised against an innovation will be both rational and emotional. Similarly the case put forward by the innovator will inevitably have strong emotional overtones. Many innovators are not easy people to live with in an organization and they are just as prone to respond emotionally to a rational argument as those opposing them. Wherever possible one must strive to resolve the conflict by objective argument. This is, however, far from easy to achieve in practice since the uncertainties surrounding so much of the information available when many of the decisions are made means that many of the judgments are inevitably highly subjective and cannot therefore be examined objectively divorced from emotional influences.

Perhaps more important than the interpersonal conflicts which arise between the innovator and other individuals is his possible conflict with the accepted norms of the organization. The climate for innovation is influenced by a number of factors, such as the rate of innovation within the industry, the company's past experience and attitude to innovation, and the age and background of top management is critical. Without their support it is unlikely that a major innovation will succeed. However, even when there is a general predisposition in favour of innovation it does not necessarily mean that top management will embrace a proposal with the enthusiasm which the innovator might expect, for its potential benefits may not be immediately apparent to them. Thus, the skill with which the technologist presents his case may have as great an influence on the outcome as several percentage points on the expected return on the investment.

Before presenting a proposal the innovator should consider carefully what objections are likely to be raised and how they might be overcome. There are two main causes of opposition arising from:

1. the nature of the project;
2. threats, real or perceived, to others within the organization.

It must be recognized that the recipients of the proposal are unlikely to be impressed by the elegance of the technological solution. Their interest lies in its merits in furthering the aims of the company, the minimization of any disruption to other activities, and the risks involved. These are the issues the innovator should address and he must be prepared to modify the proposal wherever possible to meet the anticipated objections. Consideration should be given to the following:

1. How complex is the proposal?
   - Does it involve new systems affecting people outside his normal area of influence?
   - Is it compatible with existing systems?
   - Does it have organizational implications?
   - Can it be modified to minimize these effects?
2. What is its scale of introduction?
   - Can it be tried on a pilot scale?
   - Can it be modified so that it can be introduced in a step-by-step approach?
3. What is at risk?
   - How easily can a trial be terminated and what would be the cost of failure?
4. Who is likely to be affected by the innovation, favourably or unfavourably?
   - How can allies be gained?
   - How can opposition be reduced?
   - How can the project be modified to improve its acceptance and minimize opposition?

First impressions are always important. It is better to anticipate opposition and modify a proposal in a way which makes it more attractive than to have it rejected out of hand even if it means some loss of its technological potential. In doing this it is essential to focus on its business rather than its technological merits.

Organizational and interpersonal factors are therefore of critical importance for the achievement of innovation. The innovation is not something which can be isolated from its immediate environment for the organization may itself be changed by the innovation. Consequently, technological innovation is intimately related to the management of organizational change, a topic of which companies are slowly becoming aware.

Top management plays a critical part in creating organizational attitudes favourable to innovation. The previous functional background of top management can have a significant influence on their corporate attitude to innovation, the most supportive being those with R & D or marketing experience [17]. The high proportion of technologists in top management positions in Germany and Japan suggests that this may be an important factor in their success. Where the innovator is himself a member of top management this condition is likely to arise naturally. This helps to explain why many small firms have been more successful innovators than their larger competitors. However, it is the large companies which are the greatest spenders on technology and which, because of their more complex organizational structures, need to plan consciously for the creation of a climate within which innovation can flourish. This will result from:

1. The attitudes of top management itself.
2. The skill with which the innovator presents his case.
3. An organizational structure which permits easy communication between the innovator and top management.

Commitment by one or a few individuals

'Innovations do not happen, they are made to happen.' This statement implies that behind every successful innovation there is one person or a group of people, the innovators, who are responsible for translating an idea into practice. Are we correct in this assumption that technological innovation requires a person with special attributes which are not normally regarded as essential in other managerial positions?

Analysing the requirements rationally we may conclude that the conditions for success can be assured by a systematic and methodical approach. This would be developed along the following lines:

1. Definition of a corporate strategy stating the types of business, products, and markets in which the company intends to engage.
2. Formulation of a strategy for R & D to ensure that the work of the R & D department is integrated with the corporate strategy.
3. A formal project selection system which would evaluate proposals in relation to specific financial and organizational objectives.
4. A statement of detailed project specifications against which subsequent performance can be measured.
5. Periodic evaluation of the project to assess whether the stated objectives are being met.
6. Managerial control procedures to ensure that the resources allocated to the project are being used effectively.

With such a system good management working within a formalized system should ensure the successful conduct of the project. This mechanistic approach appeals to those managers who believe that the solution of most managerial problems lies in the disciplines of planning, objective setting and control. It is difficult to refute the logic of this approach and much of the discussion in this book will be devoted to developing it further. But is this sufficient?

Under the pressure to regulate R & D expenditure there has been a trend in recent years towards increased formalization, particularly in large organizations. This has resulted in tighter financial control and has enabled many projects to be brought to a successful conclusion, particularly when the nature of the project presents no major challenge to accepted thinking within the company. There is, however, growing concern that the larger companies which have been making the maximum use of sophisticated management techniques are not always those that have been most successful in innovation. It would appear that there is an additional ingredient.

Several research studies have suggested that the demands of innovation are such that it will only succeed when driven forward by a highly committed individual who places the success of the project above all other considerations and for which he is prepared to jeopardize both his career and personal interests. This person is compared with the business entrepreneur who follows one idea with single-minded attention. He is likely to be a man of action, highly motivated, courageous, and a risk-taker. In the literature (Kanter [18], Peters [19], Pinchot [20], Roberts [21]) these individuals have been variously termed, 'project champion', 'technological entrepreneur' or 'intrapreneur'.

Why are they necessary? Organizations, which are collections of people, tend to be risk averse: they value stability and fear uncertainty. Formal systems and procedures, based on past experience, are designed to aid decision-making by delineating responsibility and enabling the familiar to be progressed more efficiently; they favour incremental change. But technical change does not necessarily fit this pattern. It is likely to require organizational change and thereby disturbs the status quo. It involves the adoption of new technologies, implying the obsolescence of the old technologies and those with expertise in them. Production equipment and manufacturing skills may become redundant. Financial risks are unavoidable. Sales personnel have to acquire knowledge of radically new products. Thus a large number of people feel threatened – some with good cause – and will react by resisting the innovation. These barriers are real and cannot be ignored. Thus they are unlikely to be overcome without the intervention of someone with the characteristics we have described. However, he will only be able to succeed in an environment which is supportive or, more cynically, is too ineffective to stop him. There are, however, a few companies – of which 3M is an outstanding example – which have structured their organizations to foster innovation and give active support to the intrapreneur.

The project champion arises naturally within an organization and is likely to be associated with radical innovations. He cannot be appointed although he can be stopped. Many of these same characteristics are still necessary for the successful completion of less ambitious projects. In this situation it is usual to appoint a project manager, whose role will be discussed in Chapter 7. This is a critical function which requires a high level of commitment and drive combined with total responsibility for all aspects of a project if it is to be effective.

It must be stressed that it is people not organizations which innovate. It is they who are creative and entrepreneurial. The organizational structures and procedures must fit their needs and harness their abilities. When the innovator is forced to subordinate his needs to existing organizational systems designed for a different purpose he is unlikely to succeed.

A production orientation

Whatever the merits of the technical innovation, it is of no value until it emerges as a product offered to the market. It must be produced and at the lowest possible cost. Technology contributes to product cost in three ways – the inherent characteristics of the product design, the choice of the production processes and their effective operation.

For organizational reasons R & D/design and production are separated, sometimes by a considerable distance. This can lead to a communication gap as serious as that between R & D and marketing. This is exacerbated by the different timescales on which they operate. R & D is concerned with the future whereas the thinking of production is the meeting of current demands.

Although the oft quoted stories of designs which cannot be made are probably apocryphal there are many examples where they are extremely and unnecessarily expensive to produce. One example is the Issigonis designed Mini car which revolutionized small car design; in spite of an exceptionally long product life of several decades it is said that it has never been profitable due to the complexity of its design. This may be exceptional but it is symptomatic of a common failing. Walleigh [22] lists the factors to be considered in overcoming these problems as: reduction of the number of parts, designing parts so that they can be assembled in only one way, simplifying the assembly process, making products easy to test, using common components over product families and relating tolerances to manufacturing capability. Gomory [23] reports the application of this approach by IBM to their Proprinter design which reduced the number of parts from 150 for an equivalent product to 62, replaced 20 parts by one moulded plastic frame, eliminated all parts requiring human adjustment and adopted a layered design so that it could be assembled by robots. Value engineering is widely used to overcome these difficulties but should be regarded as an organizational response to rectify problems which should not have been present in the original design.

There is also need to coordinate the design and production processes with the anticipated sales volume. A design which is suitable for small-scale batch production may differ considerably from what is appropriate for mass production. Such considerations should be taken into account at an early stage in the design process.

Current advances in materials technology with the increasing ability to design new materials to meet specific needs raise some interesting problems for both R & D and production. The potential to specify the properties of the materials a designer requires forces him to work more closely with materials scientists. This demands knowledge of an area with which he may be unfamiliar. Furthermore as Turner [24] states: 'These designed materials only come into existence in the process of manufacturing the product itself. In economic terms, these materials are not intermediate goods, but components of finished goods.' Similar considerations arise with the introduction of adhesives to replace the mechanical bonding of components. These advances will have a profound influence on both manufacturing technologies and investment which cannot be achieved without a close integration of production personnel in the design process. The organizational implications may be as difficult to resolve as the technical problems and may be a major determinant of the effective diffusion of these new technologies.

Many difficulties can arise when production commences. A product may have performed perfectly as a prototype where it has been produced by highly skilled and knowledgeable specialists who often carry out their own unrecorded minor modifications. Much information carried in the heads of the prototype staff may not be transferred to production. Today, with the widespread use of CAD which enables the rapid updating of drawings, there is no excuse for this. Problems also occur because the production operatives may have lower skill levels and are accustomed to working in well defined situations.

The selection of the production process is beyond the scope of this book. However, the operational improvement of these processes is or should be a responsibility of R & D. This is often hampered by the lack of a common database, the inability to install instrumentation on production equipment, and difficulty in gaining access to equipment for experimentation when it is operated continuously.

Many of these problems arise because the innovation process is seen as sequential — design to production to marketing — whereas all stages should be progressed in parallel, although the contribution of each activity will change as it proceeds. The sequential process not only leads to increased cost, it also causes delays in the introduction of a new product which may be critical if it has a short product life. In many industries today product lives are shortening; this makes it imperative that the time to market is kept to a minimum and that there are no teething problems. Whilst a close relationship between R & D, design and production has always been important, it must now be regarded as an increasingly essential element in the innovation process.

Thus there is need for:

1. An integrated approach involving R & D, design, production and marketing.
2. An organizational structure based upon the flow of activity rather than on parochial departmental interests.
3. Procedures which foster the flow of information and personnel between the technological departments and production.
4. The shortening of lead times between R & D and production.
5. An emphasis on 'getting it right first time'.
6. The fostering of mutual understanding and education in the problems facing people in the two departments.

Discussion

Factors in addition to those discussed so far may be necessary to enable projects to succeed in certain circumstances. But the body of evidence which is being built up as the result of many research studies appears to confirm that those mentioned play a critical part. Not all studies place the same weight on any particular factor but when the conclusions from a number of studies in the USA and Europe are considered together there seems strong grounds to believe, although not to prove conclusively, that not only is each factor of importance in certain innovations but that there is a high probability that all in combination are necessary.

Two research projects conducted by Langrish *et al.* and Freeman *et al.* [25, 26] in the period 1966–72 established the main factors for the commercial success of technological innovation. These are still widely quoted in the literature and the continuing relevance of their findings is confirmed by subsequent studies.

Langrish *et al.* [25] investigating 84 innovations granted Queen's Awards for technological innovation in 1966 and 1967, identified seven factors of importance to the firms' success:

1. Top person: the presence of an outstanding person in a position of authority.
2. Other person: some other type of outstanding individual.
3. Clear identification of a need.
4. The realization of the potential usefulness of a discovery.
5. Good cooperation.
6. Availability of resources.
7. Help from government sources.

and six factors which caused delay to an innovation:

1. Some other technology not sufficiently developed.
2. No market or need.
3. Potential not recognized by management.
4. Resistance to new ideas.
5. Shortage of resources.
6. Poor cooperation or communication.

Project SAPPHO under Freeman *et al.* [26] at the University of Sussex examined 29 pairs of similar projects; in each pair, one project was successful and the other less so. The report concludes:

The clear-cut differences within pairs which do fall into a consistent pattern distinguishing between success and failure can be summarized in five statements:
1. Successful innovators were seen to have a much better understanding of user needs.
2. Successful innovators pay much more attention to marketing.
3. Successful innovators perform the development work more efficiently than failures, but not necessarily more quickly.
4. Successful innovators make more effective use of outside technology and outside advice, even though they perform more of the work in-house.
5. The responsible individuals in the successful attempts are usually more senior and have greater authority than their counterparts who fail.

Cooper [27] from a study of thirty randomly selected industrial product companies identified seven different product models. The most successful approach, which he termed the 'balanced complete process', had an activity time breakdown of: marketing 31.3%, technical/production 55%, and evaluation 13.7%. The least successful were 'design dominated': marketing 28.3%, technical/production 71.7%, and evaluation 0%.

These studies focus on the management processes of innovation. However, it must not be forgotten that a precondition for success must be the quality of the product. If that is deficient none of the other factors can compensate. This is borne out by the research of Cooper and Kleinschmidt [28] into 203 new products, both successes and failures, from which they concluded that: 'product superiority is the number one factor influencing commercial success and the project definition and early pre-development activities are the most critical steps in the new product development programme.'

In recent years a number of studies have examined the characteristics of successful companies. Peters and Waterman [29] list these as:

1. A bias for action.
2. Closeness to the customer.
3. Autonomy and entrepreneurship.
4. Productivity through people.
5. Hands-on, value driven.
6. 'Stick to the knitting', i.e. stay reasonably close to the business they know.
7. Simple organizational form with a lean staff.
8. Simultaneous loose-tight properties, i.e. a combination of centralization and decentralization.

A study conducted by McKinsey [30] identified the following as distinguishing between leaders and laggards in commercializing technology, in that they:

1. Commercialize more new products and processes than their comparable competitors.
2. Incorporate several times as many technologies in their products.
3. Bring products to markets in less than half the time.
4. Compete in more different product and geographic markets.

These studies taken together lend support to the importance of the factors which we have already discussed. These can be represented diagrammatically (Fig. 1.4).

If the need for all these factors in combination is accepted then the absence of one is likely to jeopardize the successful outcome of a project. A creative idea accepted by a selection procedure may fail to reach a successful conclusion in the absence of a project champion. But a project champion committed to a creative idea may do untold harm if not linked into market and corporate needs through a formal evaluation system. Similarly, resistance from an organization opposed to innovation may smother the project even when all the other factors are present.

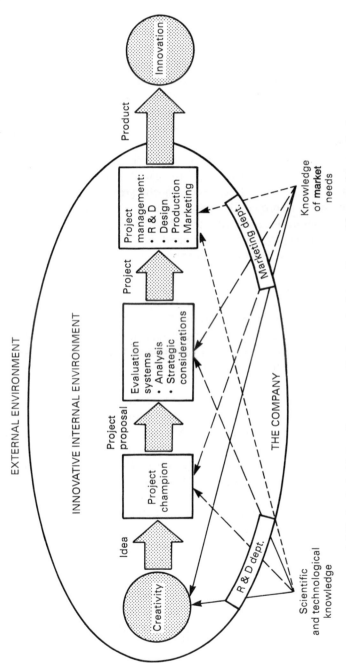

1.4 Technological innovation as a result of complex interactions — 'the egg of innovation'

To the practising manager concerned to foster innovation within his organization, analysis to show whether these factors are present may help him to understand the problem. It does not necessarily assist him in solving it. The balances involved are highly sensitive and are unlikely to be achieved by a simple mechanistic approach. For example, the realization that shortage of creative ideas is a problem is unlikely to be solved simply by the recruitment of people regarded as creative. For the condition he diagnosed is more likely to have been brought about by the nature of the organization than by chance. Unless he can identify why the organization has been unable to recruit or to retain creative people in the past and rectify this situation he is unlikely to be any more successful in the future.

Thus we see that one factor cannot be considered in isolation since they interact. Frequently they work against each other because the human characteristics required for effectiveness in one area are different from those needed in another. The project champion, because of his concern with innovation and change, comes into conflict with the administrators who are concerned with preserving the status quo. On the other hand the administrator may be appalled by the innovator's apparent disregard for organizational restraints which have been designed to eliminate the waste which his activities seem to incur. This poses a difficult dilemma for the large organization to resolve.

The attitude of the administrator is reflected in a dislike of the uncertainty and risk, inseparable from innovation. The rigid enforcement of a project selection system which demands detailed financial justification at a stage when it is not possible to give accurate forecasts can kill creativity. Taken too far this can ensure that only low-risk projects for which it is possible to make reasonably accurate estimates will be accepted. The system instead of being a filter becomes a barrier to innovation. One example which came to the author's notice was a large manufacturing company which had been a leader in the application of modern management techniques. Its management information system enabled the identification of products with low earning power which it wished to drop from the product range. Yet when replacement products were sought no suggestions for any new products could be found which offered any promise of commercial success. Over the years there had been a steady drift of the most creative people away from the company. This was attributed to the barriers which had been erected in the form of the detailed financial justifications required before approval of any new project ideas. In this case the company successfully employed a consultant who applied one of the recently developed techniques for stimulating creative problem-solving.

An interesting contrast was provided by another company in the same industry where the new product development manager expressed his difficulty in selecting products to develop from a surfeit of project proposals. The two companies had completely different perceptions of the characteristics of their industry, due largely to the balance between the creative and the evaluative emphases resulting from their organizational philosophies.

The research indicates a high degree of consensus regarding the requirements for successful technological innovation. In later chapters we shall examine what technological managers can do to satisfy these conditions.

Case histories of successful innovations

Reference has already been made to research studies where a number of successful innovations and failures have been investigated. Readers who wish to develop a depth of understanding are advised to refer to these studies. However, it is useful to draw upon two well known and outstandingly successful innovations to illustrate the importance of some of the factors which we have identified. The first of these, the development of the float glass process, dates back to the late 1950s. It is described here because it is still a classic example of an innovation which transformed an industry, it has been well documented, and it provides a fascinating illustration of the complex interplay of factors leading to success. The second, fibre optics, has been equally important commercially. Today it is a major contributor to the profits of STC which supported the pioneering efforts of innovators over twenty years ago. Both examples illustrate the long timescales required from inception to major market success. It has been suggested that in the short-term financial climate of the 1990s float glass would not have been developed. Although it is beyond the scope of this book to discuss the problems of 'short-termism' in Western business, it is worth noting that much of the success of Japanese industry has been based upon the adoption of a long-term, 20 to 25 year, time horizon for innovation.

Development of the float glass process

Glass as a manufactured material was known to the ancient Egyptians. Often centuries had elapsed between major innovations in the method of production. During the twentieth century, however, the rate of progress had accelerated although improvements were mainly concerned with increasing productivity without changing the basic manufacturing process which consisted of grinding the surface of a continuous ribbon of glass to give it the dimensional and surface finish properties required. Although productivity had been increased by a process to grind the two sides of the glass simultaneously both wastage rates and the manufacturing costs remained high.

The float glass process, which is now used by all the major manufacturers of plate glass, was an entirely new concept developed by Pilkington whereby molten glass is fed continuously on to the surface of a bath of molten tin upon which it floats. The smooth surface and constant thickness of the resulting glass completely eliminated the need for a grinding operation. Simple as was the basic concept of floating glass upon tin, the development of a practical mass production process was long and costly. There is no doubt that float glass would not have been developed but for the efforts of one man, Dr L. A. B. Pilkington. His own account [31] of the problems which had to be faced makes fascinating reading. The enthusiasm of Pilkington, the 'project champion', and his ability to transmit this enthusiasm to those working with him comes through clearly in his account where one reads of, 'one of the most exciting things in the history of the flat glass industry' where '. . . we were all tremendously excited and enthusiastic' and '. . . to be a good development man you must be a born optimist'.

Thus we can identify a creative idea and the project champion, but those together would have come to naught without the continued support of the company's top management who were required to sanction the considerable costs which were continuously escalating during the course of development. We read that '. . . the board were only interested in a cold blooded, objective analysis of the project and its progress.' Presumably, Pilkington was able to satisfy them for we read that 'In 1954 the Pilkington board decided to give the project the highest possible priority, so that success or failures would be decided as early as possible,' although they must have been well aware of the risks inherent in such a project and were prepared to accept them since he adds, 'However, at the time when the board's decision was taken it seemed quite likely that glass coming into contact with metal would always be spoiled too much to make a stable product.' At one time it looked as if their worst fears might have been realized when the project ran into serious technical difficulties: 'On our first production plant we made unusable glass for one year and two months. I had to report regularly to the board and every month put in a requisition to justify another month's expenditure of £100,000. It was a tremendous credit to the board that they gave unwavering support throughout.'

Thus one can see that the board of the company was prepared to accept risk and encourage innovation for its own sake as a business opportunity, since their existing product was under no immediate threat from a competitive innovation. But to gain their support and its continuation through a period of considerable difficulty Dr Pilkington must have been able to convince them not only of the worth of the project but of his own ability to conclude it successfully. It was perhaps also helpful in establishing good communications that he was a member of the family, albeit a distant relative, in what was at that time a privately owned company. For we may assume that he had easy and informal access to the board unhindered by the hierarchical barriers which might have obstructed another member of the company.

Uncertainty is inseparable from the development of a major innovation. It is not unusual to experience major troubles, as was the case here, when moving from pilot plant to full-scale production. At each stage of development new unknowns appear often of a type which are unlikely to have been anticipated. Referring to the time when the decision was taken to construct the first production plant Pilkington comments: 'We thought we were very much more knowledgeable than we turned out to be' and 'in retrospect we were woefully unaware of the magnitude of the problems we were going to face when we reached a mass production scale.' It is interesting to note that many of the difficulties experienced arose from the absence of a theoretical understanding of the processes involved, which followed rather than preceded the development. Technological innovation is frequently represented as a logical sequence starting with basic or fundamental research leading to applied research before development is commenced. Although this often represents what happens and ensures that the uncertainties at each stage are minimized, the higher risk approach where development is commenced before the theoretical base is established often leads to the most successful innovations as well as some of the more notable failures. This supports Roberts's conclusion that successful innovators are development rather than research orientated. Research is, of course, not to be denigrated for a theoretical understanding is important and can lead to further development of the product or process as was the case with float glass.

The development problems which were experienced and the consequent cost escalations indicate the limitations of a project selection system which places great reliance on a cost-benefit analysis. The difficulties with float glass development occurred at a late stage in development. A cautious approach might have delayed the construction of a production plant until more was known of the technology. This might well have reduced the costs associated with the delays but it is highly likely that the date of commercial launch would also have been delayed.

With the wisdom of hindsight we now know that the project was an outstanding success yielding a reduction in production costs of one-quarter and in plant size of over one-third. The cost was seven years effort and £4 million before the first saleable glass was manufactured.

What can we learn from the float glass story? First, we can see that one radical innovation transformed a major industry and the competitive position of the innovating company; and conversely, it had a serious effect upon other producers particularly when they were initially reluctant to license the new process. Secondly, the successful innovation was able to offer a significant economic advantage through a substantial reduction of manufacturing costs. Thirdly, the new idea was carried to fruition largely through the efforts of one man, the 'project champion', working within an organization which was prepared to back him. Fourthly, the uncertainties and risks were high and the difficulties experienced and their financial consequences could not have been forecast at the outset. Fifthly, the technical problems were successfully surmounted in advance of theoretical knowledge.

STC and optical fibres

In contrast to float glass, STC operates in the fast-moving electronics industry. In 1986 STC gained a £245 million contract to install the first private optical fibre transatlantic telecommunications cable. This was foreseen in the Chairman's annual report of 1982 when he reported that the company was 'at the leading edge in optical fibre technology and we have 60% of the world's available business.'

This technology is an undoubted success for STC but how did it originate? The early development of optical fibres was reported in *The Sunday Times* (14 May 1978) at a time when STC had just completed its first experimental telephone link between Hitchin and Stevenage at a cost of £500,000. In an interview the technical director, Dr Jock Marsh, traced the development back to 1966 when the inventors of optical fibre, Charles Kao and George Hockham, asked the Board for the first £10,000 research allocation. Marsh stated: 'The principle that light signals would have a higher bandwidth was well understood, but no one believed that it was time to spend large sums of money trying to make it work. It was their sheer enthusiasm which convinced us. To keep them happy we had to give them the money.'

Again we note that the original idea did not emerge from a formal analysis of requirements but from the creativity and entrepreneurial effort of two individuals within an organization that was prepared to support them in spite of severe reservations. In both cases a considerable period of time elapsed before the innovative effort was translated into business and financial success. Although formal managerial systems are essential, and are a major theme of this book, they must be applied within an understanding of the lessons we can draw from a study of what actually happens. A balance must be sought between the support of individuals, which could lead to anarchy, and formal systems which can inhibit innovation if applied without understanding. The striking of an appropriate balance between these conflicting pressures is the key to the successful management of the innovation process.

Technology and the industry life-cycle

Technological innovation can take many forms. It may be the introduction of a radically new product, the incremental improvement of existing products or the improvement of manufacturing processes. The relative value of these innovations depends upon the stage of development of the industry. It is thus important to understand how a technology-based industry evolves in order to assess how a company can exploit its potential and manage its technological activities. The concept of the industrial life-cycle provides a valuable framework for understanding the innovation process. The life cycle can be conveniently divided into five stages – incubation, technological growth and diversity, market growth and segmentation, maturity and decline (Fig. 1.5).

Stage 1 – Incubation

When a new technology or generic product type first emerges it is clouded in uncertainty. It is not known where it might be exploited, its performance is low, the reliability poor and the cost high. There is no established market and there may be considerable doubt as to whether it will satisfy a real market need. High temperature superconductivity exhibits many of these characteristics. Progress is likely to be driven by one or a few enthusiasts.

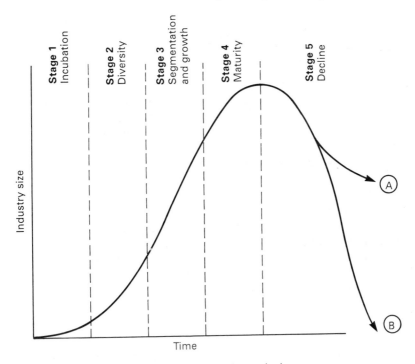

A Slow decline due to increased product longevity in
  a replacement market.
B Rapid decline due to emergence of new technology.

1.5 The industry life-cycle

If it is to find customers it is likely to be those with specialist requirements
where the performance of the technology compensates for the low reliability
and high cost. Thus the first applications of semiconductors were in aerospace
rather than in the consumer products which were to follow later. In some cases
such as composite materials the technology may languish in this stage for many
years although the underlying performance of the technology may be rising
rapidly. This stage is characterized by over-optimism accompanied by investment
for a market which usually fails to develop as quickly as anticipated, often
accompanied by a high rate of bankruptcy amongst the pioneers. Tait *et al.*
[32] have recorded this phenomenon in biotechnology in the early 1980s.

Stage 2 – Technological growth and diversity

The specialist applications of stage 1 establish the credibility of the technology
and attract a growing number of entrants into the industry. This causes a rapid
advance in the technology which is reflected in a rapid succession of products
with a significantly higher performance than their predecessors. Product lives
are short.

There is also a diversity of technological approaches before the main characteristics of a dominant design begin to emerge towards the end of this stage. Thus in the early days of the motor car there were steam and petrol engines, tiller and wheel steering, chain and shaft transmission before the main features of the car as we know it today were determined. Many companies fail because they are unable to keep pace with the rapid technological development or continue with an approach which has been overtaken, as exemplified by those who continued to support germanium rather than silicon as a semiconductor material.

Thus the main characteristics of stage 2 are:

1. Technology driven growth.
2. Short product lives.
3. Success associated with being first to market a product with a significantly higher performance.
4. High mortality of companies which:
   - fail to match the performance of their competitors;
   - support the wrong technological solution.

Stage 3 – Market growth and segmentation

With the emergence of a dominant design the industry enters a period of rapid sales growth. At the same time the rate of technological growth tends to slacken. Thus it becomes increasingly difficult to distinguish a product solely on the basis of its technological performance. As a consequence competitive advantage is more likely to be derived from designing products which meet the requirements of a particular group of customers – the market segments. In so doing the role of marketing increases and the technological emphasis is focused more on incremental improvement aimed at clearly defined market segments. Radical innovation is still possible but less frequent than previously.

Thus the characteristics of stage 3 are:

1. Rapid growth in the industry's sales volume.
2. Increasing market segmentation.
3. A lower rate of technological advance.
4. Technological innovation based more on incremental improvement.
5. A greater role for the marketing function in new product decisions.
6. A further reduction in the number of companies in the industry.

Stage 4 – Maturity

As the industry matures it becomes increasingly difficult to enhance product performance. The potential of technology has been largely exploited as it approaches its own limit (see Chapter 8). Thus many of the product improvements are little more than cosmetic. The products offered by competitors are difficult to distinguish and have many of the characteristics of a commodity. Price becomes a more important determinant in the buying decision. In this situation where it is difficult to establish a price differential profit is closely associated with minimizing the cost of production. Thus the main contribution of technology is in process innovation.

In this stage one often observes a segmentation within the industry between the producers for the mass market with a high emphasis on cost and smaller producers who specialize in satisfying niche markets. This latter strategy offers opportunity for technological innovations which add value to the product in relatively small but high margin markets.

Price is not, however, the only contributor to gaining a competitive advantage in this stage. Quality, responsiveness to the market, service and delivery all become of increasing importance. Thus it is not surprising that companies in this stage have been the first adopters of Computer Integrated Manufacturing (CIM) which enables a mass producer to satisfy small market segments economically as well as satisfying the other market needs noted above. This can have a profound effect on the organization for innovation since it demands a close integration between product design and manufacturing.

Thus the characteristics of stage 4 are:

1. The approach of market saturation with demand arising mostly for replacements.
2. The mass market supplied by a few high volume producers.
3. A very low likelihood of radical product innovation.
4. The technological emphasis of high volume producers on process technology.
5. Opportunities for smaller companies to exploit niche markets by applying technology to design high value added products.

Stage 5 – Decline

The mature stage may last a long time. However, it is likely to be terminated eventually by the emergence of a radically new technology which satisfies the same market need – synthetic for natural fibres, electronic for mechanical products, air for sea travel. History indicates that it is extremely difficult for the market leader in the old technology to make the transition to what is, in effect, a new industry. Technological expertise would appear to provide the competitive advantage rather than experience of the market. This results in 'innovation by invasion' where the invader armed with the new technology displaces the former market leader. There are a number of reasons for this related to the attitudinal and organizational factors we shall examine later. However, there is no inevitability about this and the role of management, both corporate and technical, must be to seek ways of meeting the challenge.

There are some instances where the market remains but the industry contracts due to further technological advances. A mature market consists almost entirely of replacement sales which are highly dependent upon the in-use life of the product. In the absence of new features which are rare in a mature market items will only be purchased when the old one is worn out. A company which can design a product that will last longer will gain a competitive advantage but at the expense of reducing the size of the total market. Electric batteries and car tyres are two industries which have suffered from this phenomenon. In this situation there will be many losers whose only hope of survival is to diversify into new products, markets or businesses. It is also not uncommon for the initiative for this type of innovation to be taken by a small producer who is less vulnerable to a shrinkage in the total market.

A similar effect is noted when technological advances enable the same need to be satisfied at lower cost (e.g. electronic components) or by using a smaller amount of the product (e.g. 'smart drugs').

Thus if a company is to survive stage 5 it must either:

1. Acquire expertise in the new technology and apply it effectively, building on its existing market strength.
2. Use its existing technology to enter new markets.
3. Aim to dominate the replacement market through using its technology to increase the in-use life of the product, recognizing that this is likely also to be the objective of its main competitors.

In all these cases a proactive policy is essential if timely action is to be taken.

Discussion of the industry life-cycle

The industry life-cycle is a useful concept in aiding one's understanding of the technological innovation process, although not all industries evolve as simply as the cycle might suggest. It must be stressed that in discussing the role of technology we have been considering where the greatest potential gains might be achieved. Thus it should not be inferred that process technology has no part to play in the early stages or that radical innovation is impossible in the later stages. However, examination of the current position of an industry on the cycle and where it is progressing can be of great value in determining a technological strategy.

It can also be seen that the relative status of the functions changes as the cycle evolves. In the first two stages the technological contribution is central to business success and it is appropriate that the R & D director is pre-eminent amongst his peers. This changes first to marketing and finally to finance and production. Of course, they are all important at all times but the cycle indicates where the company is likely to gain its greatest competitive advantage.

The style and system of management will also evolve. In the early stages entrepreneurship, informal systems and a focus on the timely introduction of new products are more important than the minimization of development cost. Too much emphasis on cost control might introduce delays which remove the competitive advantage the product might possess if launched early. As one progresses up the cycle cost becomes more important as do the systems by which it is controlled. This is accompanied by a formalizatoin of management systems. Thus one can conclude that there is no 'right' system for managing technological innovation but only one that is appropriate to the needs of the time.

These considerations are of great significance in managing the innovation process. It can be seen that in a multi-technology company it may be necessary to adopt different management systems in its various parts. Furthermore, it is not easy for managements to accept that new innovative ventures need to be managed in a loose informal way which runs counter to the company's formal procedures which have been developed over the years. This, more than any other factor, explains why so many radical innovations are introduced by new ventures rather than the established industry leaders. These problems can be overcome and ways in which it can be achieved are discussed in later chapters.

These implications of the industry life cycle are summarized in Table 1.1.

Table 1.1 *The industry life-cycle and its implications*

	Stage 1	*Stage 2*	*Stage 3*	*Stage 4*	*Stage 5*
Technology emphasis	Invention Applied research Radical innovation	Product performance Speed of development	Dominant design Fewer new products	Process innovation Minor improvement	In-use life Technological diversification
Market emphasis	Specialist Very small	Short product lives High variety	Rapid growth Segmentation	Price Promotion Competition	Price Quality Service
High status	R & D	R & D and marketing	Marketing	Production Finance	Production Finance Marketing
Cost emphasis	Low	Low	Increasing	High	Very high
Organization	Informal	Informal	Formalizing	Formal	Formal
Risk and uncertainty	Very high	High	Low	Medium	High

Electronic components and the industry life-cycle

Table 1.2 shows the top ten world manufacturers in one industry for the period 1955–88. At the beginning of this period the transistor was beginning to substitute for the thermionic valve. It will be noted that the top valve manufacturers were, and several still are, major electronic companies. A few of them initially invested in the successor technologies although none became a leader and none remained in the top ten by 1975. Thus they lost a valuable market. For a variety of reasons they were unable or unwilling to exploit the new technology. RCA, for example, placed transistor development within the valve department where psychological and financial investment in valves

Table 1.2 *Electronic components and the industrial life-cycle – Top Ten companies, 1955–88*

	Valves	Transistors	Semiconductors							
	1955	1955	1960	1965	1975	1980	1983	1984	1985	1988
RCA	1	7	5	6	8	–	–	–	–	–
Sylvania	2	4	10	–	–	–	–	–	–	–
GE	3	6	4	5	–	–	–	–	–	–
Raytheon	4	–	–	10	–	–	–	–	–	–
Westinghouse	5	8	–	–	–	–	–	–	–	–
Amperex	6	–	–	–	–	–	–	–	–	–
National Video	7	–	–	–	–	–	–	–	–	–
Ranland	8	–	–	–	–	–	–	–	–	–
Eimac	9	–	–	–	–	–	–	–	–	–
Lansdale Tubes	10	–	–	–	–	–	–	–	–	–
Hughes	–	1	9	–	–	–	–	–	–	–
Transitron	–	2	2	9	–	–	–	–	–	–
Philco	–	3	3	8	–	–	–	–	–	–
Texas Inst.	–	5	1	1	1	1	1	1	2	5
Motorola	–	9	6	2	5	3	2	3	4	4
Clevite	–	10	7	–	–	–	–	–	–	–
Fairchild	–	–	8	3	2	7	–	–	–	–
GI	–	–	–	4	7	–	–	–	–	–
Sprague	–	–	–	7	–	–	–	–	–	–
National Semi-Conductor	–	–	–	–	3	2	7	6	8	–
Intel	–	–	–	–	4	5	8	7	7	10
Rockwell	–	–	–	–	6	–	–	–	–	–
Signetics	–	–	–	–	9	4	6	9	10	6
AMI	–	–	–	–	10	–	–	–	–	–
NEC	–	–	–	–	–	6	3	2	1	1
Hitachi	–	–	–	–	–	8	4	4	3	2
Fugitsu	–	–	–	–	–	9	9	8	6	7
Toshiba	–	–	–	–	–	10	5	5	5	3
Mostek	–	–	–	–	–	10	–	–	–	–
Matsushita	–	–	–	–	–	–	10	10	9	8
Mitsubishi	–	–	–	–	–	–	–	–	–	9

provided an unfavourable corporate climate for the new technology.

The successful companies were mostly new entrepreneurial ventures which grew rapidly as the new technology advanced. However, with the notable exceptions of Texas Instruments and Motorola, these also lost position in the period 1965–75. In some cases this was because they continued to support germanium when silicon was becoming the dominant technology; in others it was due to the rigidities which evolved as the companies grew in size. Thus one sees the emergence of a new wave of entrepreneurial companies, often referred to as the 'Fairchildren' due to the number which were founded by former Fairchild employees.

However, as the technology matured in the 1980s the emphasis changed to production technology and the sophisticated managerial systems characteristic of the large companies, in this case Japanese.

Thus one sees the emergence of new industry leaders at each stage of the life-cycle reflecting the inability of the previous leaders to manage the innovation process in a way appropriate to that stage. This phenomenon is also noted in other industries although not so dramatically. However, two companies were exceptions, Texas Instruments and Motorola. Their management systems enabled them to make the necessary adjustments to succeed throughout the period, albeit losing some market share to the Japanese in recent years. This is a tribute to their managements and indicates that it is possible to manage the innovation process throughout the life-cycle. Such companies are the exceptions.

Summary

Throughout any consideration of technological innovation we are brought face to face with a dilemma — the need to control an expensive activity while at the same time providing an environment within which individual creativity and entrepreneurial drive can flourish. Somehow these conflicting claims must be reconciled. If we are to attempt to manage this activity professionally it is necessary to derive guidelines for decision-making without losing sight of the inherent uncertainties and risks associated with it.

Of one thing we can be certain. Formal management techniques cannot alone guarantee success. Innovation is part science and part art. We cannot train an artist to produce great masterpieces nor program a computer to do so; yet the teaching of the elements of composition, draughtsmanship and colour balance are not considered irrelevant even when it is recognized that occasionally masterpieces are created by breaking all the rules; so with technological innovation.

Examination of research studies enables the identification of certain factors present in a considerable number of successful innovations and frequently absent in failures — a market orientation, relationship to corporate objectives, evaluation techniques, good project management, creativity, an innovative environment and a project champion. The importance of technological innovation is so critical for many businesses that it cannot be left entirely to chance and a conscious effort must be made to ensure the presence of the conditions for success. Management cannot abdicate from its responsibilities merely because of the absence of sufficient data to substantiate its decisions in advance.

Decisions then must be made. They will be better decisions if they are made with an understanding of the processes at work and within a conceptual framework. The following chapters will attempt to develop such a conceptual framework based on what research evidence is available. This will not present the reader with ready-made solutions to his problems, but it is hoped that it will enable him to fit together some of the pieces of this multi-dimensional jigsaw. For it is worth repeating what has been said earlier — while we cannot ensure success there is ample scope for improving our ability to avoid failure.

References

1. Morbey, G. K. 'R & D: Its Relationship to Company Performance', *Journal of Product Innovation Management*, **5**, No. 3, Sept. 1988.
2. Goodridge, M. and Twiss, B. C. *Management Development and Technological Innovation in Japan*, Manpower Services Commission Report No. MC28, 1986.
3. Price, Derek de S. 'The Science/Technology Relationship, the Craft of Experimental Science, and Policy for the Improvement of High Technology Innovation', *Research Policy*, **13**, No. 1, Feb. 1984.
4. Wind, Y. and Mahagan, V. 'New Product Development Process: A Perspective for Re-examination', *Journal of Product Innovation Management*, **5**, No. 4, Dec. 1988.
5. Bright, James R. *Some Management Lessons from Technological Innovation Research*, National Conference on Management of Technological Innovation, University of Bradford Management Centre, 1968.
6. *Technological Innovation: Its Environment and Management*, US Dept of Commerce, 1967.
7. Marsh, P. 'The Ideas Engine which Drives Japan', *Financial Times*, 29 May 1987.
8. Lucas, G. H. and Bush, A. J. 'The Marketing−R & D Interface: Do Personality Factors Have an Impact?', *Journal of Product Innovation Management*, **5**, No. 4, Dec. 1988.
9. Elliott, K. and Margerison, C. *Affective Dissonance Amongst Professional Personnel*, unpublished research report, University of Bradford, 1970.
10. Rosenau, M. D. 'Schedule Emphasis of New Product Development Personnel', *Journal of Product Innovation Management*, **6**, No. 4, Dec. 1989.
11. von Hippel, E. *The Sources of Innovation*, Oxford University Press, 1988.
12. Habermeir, K. F. *Product Use and Product Improvement Research Policy*, **19**, No. 3, June 1990.
13. von Hippel, E. *The User's Role in Industrial Innovation*, TIMS Studies in Management Sciences 15, 1980.
14. von Hippel, E. 'Successful Industrial Products from Customer Ideas', *Journal of Marketing*, Jan. 1978.
15. von Hippel, E. 'Transferring Process Equipment Innovations from User-Innovators to Equipment Manufacturing Firms', *R & D Management*, **8**, No. 1, Jan. 1977.
16. Schein, Edgar H. *Organizational Psychology*, Prentice-Hall, 1965.
17. Hegarty, W. H. and Hoffman, R. C. 'Product Market Innovations: A Study of Top Management Involvement Among Four Cultures', *Journal of Product Innovation Management*, **7**, No. 3, Sept. 1990.

18. Kanter, R. M. *The Change Masters: Corporate Entrepreneurs at Work*, George Allen & Unwin, 1984.
19. Peters, T. *Thriving on Chaos*, Macmillan, 1988.
20. Pinchot, G. *Intrapreneurship*, Harper & Row, 1985.
21. Roberts, E. B. *Entrepreneurs in High Technology*, Oxford University Press, 1991.
22. Walleigh, R. 'Product-Design for Low-Cost Manufacturing', *Journal of Business Strategy*, July–Aug. 1989.
23. Gomory, R. E. 'From the "Ladder of Science" to the Product Development Cycle', *Harvard Business Review*, Nov.–Dec. 1989.
24. Turner, C. *et al.* 'Materials: A New Revolutionary Generic Technology? Conditions and Policies for Innovation', *Technology Analysis and Strategic Management*, **2**, No. 3, 1990.
25. Langrish, *et al.* 'Wealth from Knowledge', *A Study of Innovation in Industry*, Macmillan, 1972.
26. Freeman, *et al. Success and Failure in Industrial Innovation*, Centre for the Study of Industrial Innovation, University of Sussex, 1972.
27. Cooper, R. G. 'The New Product Process: an Empirically Based Classification Scheme', *R & D Management*, **13**, No. 1, Jan. 1983.
28. Cooper R. G. and Kleinschmidt, E. J. 'New Products: What Separates Winners from Losers', *Journal of Product Innovation Management*, **4**, No. 3, Sept. 1987.
29. Peters, T. J. and Waterman, R. H. *In Search of Excellence*, Harper & Row, 1982.
30. Nevens, T. M. *et al.* 'Commercializing Technology: What the Best Companies Do', *Harvard Business Review*, May–June 1990 (Report of McKinsey study).
31. Pilkington, L. A. B. 'The Float Glass Process', *Proceedings of the Royal Society London*, **A314**, 1–25, 1969.
32. Tait, J. *et al.* 'The Status of Biotechnology-based Innovations', *Technology Analysis and Strategic Management*, **2**, No. 3, 1990.

Additional references

Bonnet, D. C. L. 'Nature of the R & D/Marketing Cooperation in the Design of Technologically Advanced New Industrial Products', *R & D Management*, **16**, No. 2, April 1986.

Hawthorne, E. P. *The Management of Technology*, McGraw-Hill, 1978.

Hill, C. T. and Utterback, J. M. *Technological Innovation for a Dynamic Economy*, Pergamon Policy Series, 1979.

Kay, N. M. *The Innovating Firm: A Behavioural Theory of Corporate R & D*, Macmillan, 1979.

Maidique, M. A. 'Entrepreneurs, Champions and Technological Innovation', *Sloan Management Review*, Winter 1980.

Moenaert, R. K. and Souder, W. E. 'An Analysis of the Use of Extrafunctional Information by R & D and Marketing Personnel: Review and Model', *Journal of Product Innovation Management*, 7, No. 3, Sept. 1990.

Roberts, E. B. (ed.). *Generating Technological Innovation*, Oxford University Press, 1987.

Roman, D. D. *Science, Technology and Innovation: A Systems Approach*, Gird Publishing, 1980.

Souder, W. E. 'Improving Productivity Through Technology Research', *Technology Management*, Mar./Apr. 1989.
The Conditions for Success in Technological Innovations, OECD, 1971.

Tushman, M. L. and Moore, W. L. (eds). *Readings in the Management of Innovation*, 2nd edn, Ballinger, 1988.

2

Strategies for research and development

At the moment, we should consider whether the advantages of a consciously considered strategy are worth the effort it obviously requires. Four considerations suggest an affirmative answer. They are the inadequacy of stating goals only in terms of maximum profit, the necessity of planning ahead in undertakings with long lead times, the need of influencing rather than merely responding to environmental change, and the utility of setting visible goals as an inspiration to organizational effort.

Edmund Learned

The discussions in Chapter 1 focused on an examination of what research has shown to be the main features of the innovation process likely to lead to success or failure. The challenge is to learn from these experiences and develop managerial systems which enhance the probability of a successful outcome. There are no easy answers. It will already be apparent that R & D management faces a dilemma. There is a clear need for a systematic approach based upon rational analysis and logical thought. But this must be combined with the development of an environment which nurtures the creativity and entrepreneurship which challenges the accepted wisdom.

In this chapter we shall examine the strategic aspects of innovation management whereby rational analysis is used to align the allocation of R & D resources to projects which help to achieve the stated objectives and strategies of the company. In Chapter 3 we shall explore the role of creativity in developing ideas which may not only further that aim but may also suggest entirely different opportunities for the exploitation of technology. The importance of a clearly enunciated R & D strategy has become widely recognized in recent years. However, it must be appreciated that the value of strategic thinking lies in the way it is applied rather than in the formality of the procedures. It is of little value if it merely leads to a formalization of the status quo, which is all too often the case. To make an effective contribution it must provide a rational basis for indicating the desirability of a technology-based strategic change.

A brief description of the elements of a corporate strategy will now be outlined before we explore the development of an R & D strategy.

The role of a corporate strategy

One of the distinguishing features of professional management is the effort devoted to planning and control so that all the activities of an organization can work towards a common set of objectives against which their performance can be measured. The competitive pressure of recent years, coupled with the emergence of a few very large companies in many industries and the accelerating rate of change, created an environment where top management thinking turned to the consideration of planning in relation to the total company operation. Formal corporate planning systems were initially confined to large companies; they have now gained increasing acceptance by medium-sized and smaller companies. The works of writers such as Ansoff [1] and Porter [2] are now widely read by practising managers.

The process of corporate planning involves the systematic examination of a number of interrelated elements (Fig. 2.1) which results in an explicit statement of company objectives and how they are to be achieved. The terminology used by writers in this field can be confusing to readers since there are no universally accepted definitions for such terms as objectives, aims, goals, strategy and policy. In spite of this semantic confusion there is, however, a wide measure of agreement concerning the elements involved in the process. The meanings attached to the elements shown in Fig. 2.1 are as follows:

Objectives

In this context the word 'objective' is used to define medium- to long-term specific aims for the whole organization. A clear distinction is drawn between general objectives, sometimes referred to as the purpose of the business, expressed in statements such as 'We aim to be the market leader in synthetic fibres', and specific objectives. Although generalizations may be helpful in describing the aspirations of a business they provide little guidance to managers responsible for making operational decisions. The decision-maker requires a clear statement of objectives which will usually include a definition of the company's products and markets together with quantified targets, related to a specific timescale, for such items as market share, turnover and profit. In this sense the word 'objectives' defines the end-point to be achieved (WHERE) at a specified time (WHEN).

Strategy

There are a number of paths by which any set of objectives can be reached; these are called strategies. A company which has set itself as one objective '50% growth in turnover and 60% in profit by the end of three years', has a variety of alternatives open to it:

1. Growth of market share from existing products.

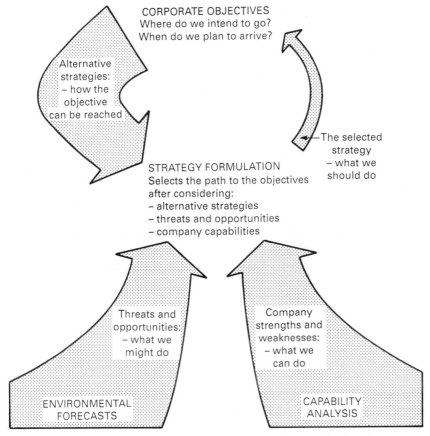

CORPORATE OBJECTIVES
Where do we intend to go?
When do we plan to arrive?

Alternative
strategies:
– how the
objective
can be reached

The selected
strategy
– what we
should do

STRATEGY FORMULATION
Selects the path to the objectives
after considering:
– alternative strategies
– threats and opportunities
– company capabilities

Threats and
opportunities:
– what we
might do

Company
strengths and
weaknesses:
– what we
can do

ENVIRONMENTAL
FORECASTS

CAPABILITY
ANALYSIS

Note: In practice this is an iterative process. It may, for instance, be found
necessary to change the objectives because they are shown to be
unrealistic when considered in relation to possible strategies.

2.1 The elements of a corporate strategy

2. Extension of the market by enlarging the product range.
3. Entering new markets, perhaps by becoming international.
4. Reduction of manufacturing costs.
5. Enlargement of the business through vertical integration without changing
   the volume of the end-product sold.
6. Growth by acquisition or merger.

and so on.

Since these choices are not mutually exclusive they give rise to a large number
of possible combinations or alternative strategies, one of which must be selected
as the most promising. The selection of which path is to be followed will be
referred to as 'strategy formulation' (HOW).

The business environment

Planning decisions are directed towards the future, a future when conditions may be very different from those at the time when the plan is drawn up. If corporate objectives are to be met they must be realistic in relation to the business environment existing at some time in the future. The same is true for the strategy. Thus a conscious view of the future, however imprecise, must be built into the planning process. Forecasts are needed, covering the time to the planning horizon, for every factor which is liable to change and to alter in some way that might affect the accomplishment of the plan. A wide range of economic, social, technological and political factors have to be taken into account. It is a lucky manager indeed who cannot identify under one of these headings some change which is going to have a substantial influence on his business during the next ten to fifteen years.

But it is not necessary to examine forecasting techniques to see that a study of the future is of great value if it enables the identification of environmental trends which may present either a threat to the existing business or an opportunity for new business, and consequently lead to the consideration of possible strategies (WHAT IT *MIGHT* DO).

Analysis of company capability

So far we have been looking at considerations which are independent of the business itself. There is no reason, apart from the difficulties inherent in forecasting, why several companies in the same industry should not identify the same threats, opportunities and possible strategies. But the ability of the business to meet the threats or exploit the opportunities depends upon a variety of factors unique to the particular company. Thus two similar companies will choose different strategies which are matched to their own capabilities. To do this it is necessary to carry out a critical evaluation of the company's strengths and weaknesses. Where marketing expertise is high one might expect a growth strategy based on increased market share or expanded markets to be favoured; R & D strength might suggest growth based upon new products; the ability to manufacture at low cost could support vertical integration; whereas financial competence and resources could be exploited through a policy of acquisition or mergers. Thus by comparing the environmental forecasts − WHAT the company *MIGHT* do − with an analysis of its capabilities − WHAT IT *CAN* DO − it is possible to make the final choice of strategy − WHAT IT *SHOULD* DO.

Strategy formulation in practice is very much an iterative process. Realistic corporate objectives cannot be set until some consideration has been given to possible strategies, forecasts and capability analysis and vice versa. Nevertheless as the planning process progresses a coherent pattern of objectives and strategies will develop which are consistent with the information available at the time.

Theoretically they will be out of date as soon as new information becomes available, although in practice major changes will only be needed infrequently. Without a high degree of stability in objectives and strategies operational planning would become extremely difficult. Nevertheless, changes will be needed from time to time. One of the dangers to be guarded against in any formal planning system is an inflexibility stemming from a reluctance to amend a plan which has absorbed a great deal of managerial time and commitment.

Strategy formulation and planning are meaningless unless they are reflected in decision-making at all levels of the organization. They will usually be promulgated through statements of key policies and action plans but the existence of a formal planning system does not by itself ensure tht the strategy will be implemented. Formal planning has become a feature of corporate life with which the R & D manager must learn to live. But implementation of the plan will rest heavily upon the new products or processes which emerge from the R & D department. Consequently, the projects selected today will determine what new products and processes are available in a few years time, and therefore what strategies are feasible for the company at that time. If these decisions do not take into account the corporate strategy the planning process becomes meaningless since the decisions which really count are being made within the R & D department.

In this context it is essential to assess the totality of the innovation system (see Chapter 4), for technological superiority does not necessarily guarantee commercial success. Sony's Betamax video recording system was regarded as technically superior to VHS but the latter captured the market and became the industry standard due to the availability of software. It is essential that a technological strategy is not developed in isolation from all those factors which will ultimately determine its successful implementation within the company and its acceptance in the market.

We shall now consider how R & D can be integrated into corporate planning by means of an R & D strategy. The discussion will focus on the formal aspects of the problem, inevitably a mechanistic approach; but it should not be forgotten that without the right managerial attitudes towards it the system merely provides a framework which may mean very little. Thus before looking at R & D strategy we shall examine how provision can be made for the unplannable − the project proposal or innovation which could not be foreseen when the plan was drawn up.

R & D as a business

A major objection to the application of formal planning to research and development is that many of the most important technological innovations originate in a random fashion. Chance plays an important role and the literature frequently alludes to serendipity − 'the faculty of making happy and unexpected discoveries by accident' (OED). But no company is going to invest heavily in technology solely as an act of faith in the hope that by backing the right people 'something will turn up'. On the other hand, it would be a short-sighted management which was not prepared to consider an unexpected innovation on the grounds that it had not been foreseen in the plans.

It is also argued that the technologist's spirit of enquiry, particularly when he is working towards the research end of the R & D spectrum, must be given some satisfaction. He should be given an opportunity to devote some of his effort to working on projects which may not appear immediately relevant to the company's needs. The argument runs that without this freedom to follow some of his own interests, the laboratory would become uncreative and it would be difficult to attract or retain high calibre technologists. Some managers would support provision of personal research on these grounds alone, irrespective of the possibility of any commercial return. These considerations arise mainly in large laboratories where applied research forms a high proportion of the total activity.

Both serendipity and the need for personal research cannot be ignored. But their existence does not destroy the rationale for planning if the planning system is sufficiently flexible to accept some activities not directed towards clearly identified ends. This can be accommodated by the recognition that a company which is investing heavily in technology is in reality engaged in two businesses:

1. The primary business defined by its corporate objectives which is directed towards the satisfaction of identified market needs.
2. A secondary technological business which is generating technology of a commercial value but often unrelated to the corporate objectives. This value will normally be realized only by selling the technology itself although in exceptional circumstances it may warrant full development of a product and the establishment of a separate manufacturing and marketing operation as a diversification.

This 'two-business' concept is illustrated schematically in Fig. 2.2. The primary business allocates financial resources to the R & D department, the majority of which are allocated to a number of projects (Nos 1 to 10 in Fig. 2.2) clearly related to the furtherance of the corporate strategy. However, a proportion of this money is set aside to be used at the discretion of the research director to support personal research or to follow lines of investigation which appear promising. Some laboratories allocate about 10% of their resources to this type of work. This could mean allowing each technologist to devote 10% of his time to personal research, but in practice it usually involves a smaller number of people working temporarily in this way for a high proportion of their time. Motivation exists for all the technologists even when they are not doing personal work themselves since they are aware that the opportunity exists. The exact proportion of funds which should be spent in this way is a matter of judgment and is a corporate decision. Once the allocation is agreed the expenditures can be budgeted and controlled. The research director is responsible for the allocation to projects, but not for deciding the proportion of his total budget to be spent in this way.

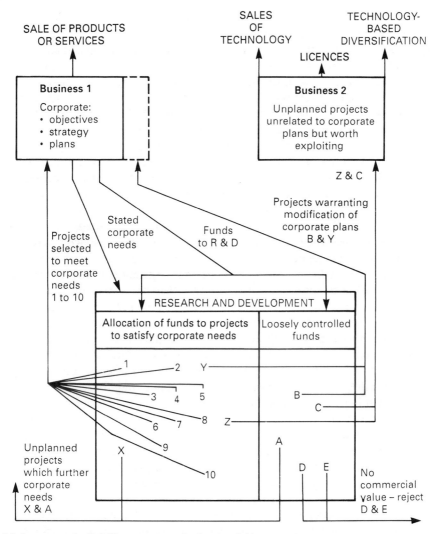

2.2 Investment in R & D supports two business activities

Referring again to Fig. 2.2, we see eight actual or potential projects which do not form part of the formal strategy-oriented programme:

1. X, Y and Z arise from the formal programme as either logical extensions of existing projects or entirely unrelated serendipitous discoveries.
2. A, B, C, D and E derive from personal research.

These eight projects may still be brought into a system of evaluation as follows:

1. *X and A*. Although these were unplanned they are seen to relate closely to

the formal programme and can be incorporated in it should they be considered to have sufficient potential.

2. *Y and B* do not fit the plan, yet they are not completely unrelated and appear highly promising. Thus we have the emergence of new information which may warrant a revision of the plan, perhaps a change in strategy or even in the corporate objectives. Where planning is inflexible projects such as these may easily be rejected for they imply change which we have already seen may arouse strong organizational resistance, not least of all from the corporate planners. One advantage of a formal system of analysis is that it brings into the open the basis of disagreement and provides a structure for objective discussion. It enables the true implications of such projects to be examined. Promising projects are likely to arise only occasionally but they may also be the most significant for the company in the long term. Their survival may depend upon the efforts of a project champion but his task should be easier when the areas of disagreement are stated explicitly in relation to a planning system.

3. *Z and C* neither fit the existing plan nor do they appear to justify changing it. But this does not necessarily mean that they have no commercial value. In another company or sometimes in a different industry they may have considerable potential. Thus the projects themselves are a marketable product which can be exploited, though they may require additional development before they can be sold. This gives rise to an additional business activity – the selling of unwanted technology. Many companies neglect these business opportunities; they take out protective patents and forget about them. Commenting on one aspect of this problem Reginald Whitaker, Deputy Chairman of the National Institute of Licensing Practitioners [3] comments: 'Any reasonably successful business has two things to sell – what it makes and how it makes it. But the majority of firms simply do not think in terms of selling their "know how". Yet this is often the turning point between a moderately successful company and an extremely profitable one.' Some companies are now awakening to this untapped source of wealth. US General Electric, for example, has a New Business Development Operation, one function of which is to market patents; for an annual fee customers receive a bimonthly portfolio of brief business ideas, further particulars of which can be obtained on payment of an additional fee.

   Once this policy is accepted new considerations arise. It is one thing to sell off unwanted technology as an opportunist venture, but, having accepted the principles, it may seem logical to extend it and deliberately channel additional funds into developing new technology for the sole purpose of sale. There may be nothing wrong in this provided it is a carefully thought out corporate decision as distinct from the uncontrolled result of entrepreneurial efforts by the research director. However, whatever the extent to which this activity is allowed to grow, the concept of two businesses provides for the maximum exploitation of the results of the investment in technology.

4. *D and E*. These two projects appear to have no commercial application and

should therefore be terminated before they reach a stage where they are absorbing additional funds and technological effort.

The line of reasoning we have followed enables the requirements of corporate planning to be reconciled with the realities of the R & D situation. While provision has been made for the development and commercialization of the unexpected, it is still brought within the bounds of corporate control.

Resource allocation to research and development

Involvement in technology

One of the most difficult problems facing top management is to decide how much it should spend on R & D. There will always be limited investment funds available and these must be divided between areas where they might be expected to give a quick return, such as advertising, and investments for the long term such as R & D. But R & D is not alone in claiming funds which will not give an immediate return, although it is the most difficult to justify on a cost:benefit basis, for it is rarely possible to correlate R & D expenditure directly with a measurable change in profitability since the effects are usually obscured by other factors. A product may, for example, fail in the market-place through deficiencies in marketing or over-pricing, although technically it might have been a success. Conversely, good merchandizing may largely counteract the deficiencies of a mediocre product. While it is difficult to isolate the R & D contribution to product profitability, it is even more difficult to correlate company profitability over a period covering such disturbances as tax changes and economic cycles, with earlier investment in R & D.

A major involvement in R & D must be a strategic decision which will be based on the judgment of top management influenced by their own value system. Some companies deliberately avoid spending more than the bare minimum on technology. This decision may form part of a coherent strategy based on a policy of avoiding the risks inherent in technological innovation and preferring to exploit their expertise in other functions such as marketing, manufacturing or finance. Other companies will commit their future to the attainment or retention of a technological lead. Both these strategies can be feasible alternatives as we shall see later. Only rarely do companies change their general level of support for technology and when they do so it often results from the installation of a new management team following a merger or takeover. Of greater concern to the R & D director is the size of his budget allocation year by year, given that there is no significant shift in the company's attitude to technological innovation.

Setting the R & D budget

The setting of the annual budget might be expected to be linked to the company's long-term corporate objectives; in practice, however, it is frequently subject to short-term fluctuations which relate more to the availability of funds than to technological needs. That this is so stems largely from the difficulty of establishing a basis for allocation which is acceptable to all parties. There are a number of approaches to the problem:

1. Interfirm comparisons.
2. A fixed relationship to turnover.
3. A fixed relationship to profits.
4. Reference to previous levels of expenditure.
5. Costing of an agreed programme.
6. Internal customer–contractor relationship.

Interfirm comparisons
Although the level of R & D expenditure varies widely between industries, it might be expected that competitive companies within an industry would be investing similar amounts if they are to retain their competitive positions. A rough guide to the order of R & D investment for a particular company should then be obtainable from an examination of its competitors' expenditure. R & D investment as a percentage of turnover is the measure frequently used. Meaningful information is, however, rarely contained in published sources. Even when it is available it may be misleading unless interpreted intelligently.

Definitions of R & D vary widely, and one company will include under this heading expenditure which another would not; thus it is unlikely that published information will be strictly comparable. Similarly, the activities of two companies are never exactly identical even when many products are directly competitive. For example, one might consider that comparison between two companies such as Philips NV and GEC plc might be useful. Yet on closer examination it becomes apparent that, not only is their product mix different, but that different products, by their very nature, are more or less R & D intensive – ranging from technologically complex products such as radar sets to electric light bulbs where the R & D content is negligible. A global figure for turnover masks the varying proportions of the different type of product in total sales and consequently the technological content in the turnover.

In spite of these objections it is still possible to make a fairly accurate estimate of a competitor's activity, provided that the available information is interpreted intelligently. The legitimate industrial intelligence which all companies should undertake reveals reliable information on such factors as the numbers employed in the laboratories from which an approximate estimate of the annual cost can be calculated. Early warning of future changes in competitive position might also be deduced from noting trends in recruitment, etc. Thus, by analysing what major competitors are doing, it is possible to obtain a rough indication of what one's own investment might have to be in order to remain competitive, although it is unlikely that simple percentages calculated from balance sheet information will be other than misleading.

A fixed relationship to turnover

Relating the R & D budget to turnover as a constant percentage is the method most frequently used. Turnover is not usually subject to violent oscillations from one year to another so this basis does ensure that the budget allocations are reasonably stable and grow in line with the size of the company. This suffers from the criticism that it refers to the present turnover of the company resulting from past investment rather than to the future to which current R & D expenditure relates.

A fixed relationship to profit

Tying the R & D budget to company profitability is highly undesirable. It implies that R & D is a luxury which can only be afforded when the company is doing well. This argument might appeal to the company finance director faced with cash-flow difficulties. He would point out that survival in the short term is his highest priority and in periods of financial stringency he must reduce expenditures which can be curtailed easily – advertising, education, R & D. But whereas advertising and education can be reinstated easily, it may take several years to rebuild a decimated laboratory.

While some understanding can be shown for the financial argument it should be opposed except in circumstances of serious emergency. Low current profitability may stem largely from uncompetitive products as a result of insufficient past investment on new product development. This should suggest an increase rather than a decrease in the R & D budget for there is no logic in sacrificing the future because of poor past performance.

It might be said, with some justification, that no company deliberately links R & D with profits. Yet the savage cuts imposed by many UK companies during periods of economic recession indicate clearly that this happens all too often in practice.

Reference to previous levels of allocation

In the absence of any other criteria the budget often results from bargaining between the research director and top management. A starting point is likely to be the previous allocation plus a margin to cover inflation, expansion, new capital equipment, etc.

Costing of an agreed programme

The research director is primarily concerned with funding for specific projects as well as maintaining the level of R & D activity. He may thus arrive at a budget by adding together the requirements for the individual items within his programme. Almost invariably, his total will exceed what he is likely to receive by a comfortable margin. Somehow his demands and likely supply must be reconciled.

Internal customer–contractor relationship

In some large multinational companies the individual business units pay for the research done on their behalf. The central research laboratory then assumes some of the characteristics of an external contract research organization. In addition it is essential to make provision for building the knowledge base and for long-term applied research which will ultimately be of benefit to the company as a whole. For example, in the Royal Dutch Shell Group the total R & D budget is set by the Managing Director, within which 10–12% is allocated to long-term research. Customer companies pay for specific projects but 25% of this contribution is allocated to exploratory research. However, the overall programme is reviewed centrally to ensure that the customer demands are kept within the agreed total budget.

Discussion

Judgment and negotiation will inevitably play an important part in the allocation of resources to R & D(Kay [4]). No single method provides a satisfactory basis. Consideration should, however, be given to the following:

1. Expenditure by competitors.
2. The adequacy of previous allocations in relation to the needs of the R & D strategy.
3. Company long-term growth objectives. (This relationship is explored at greater length under *gap analysis* – see p. 55.)
4. The need for stability. Whereas contractions are painful in a labour-intensive activity, rapid expansions are also difficult to effect smoothly. This argues for a steady rate of growth, year by year, without violent fluctuations.
5. Distortions introduced by large projects. Although every effort should be made to programme projects so as to provide a steady volume of work, there will be occasions when one or more major projects place heavy demands on the budget in a particular year. It is generally more desirable to increase the budget temporarily than to hold back important programmes.

Occasional budget cuts, provided they are not too vicious, can, however, sometimes have a salutary effect. Due to the impossibility of measuring R & D effectiveness, unproductive equipment and people can easily accumulate unnoticed, particularly after a period of steady growth.

R & D strategy in the decision-making process

If corporate planning is to influence R & D policy its main effect will be seen through decisions relating to the selection of projects and the allocation of resources between them. A system is needed whereby the whole of the decision-making process from the top management down to R & D management can be integrated into a coherent pattern for operational decisions. However, within the R & D department, decisions are commonly taken on a project to project basis where individual projects are considered on their own merits with little reference to their contribution to a balanced portfolio. Thus the criteria applied in project selection are evaluated as if the project had intrinsic merits of its own independent of the organization within which it will be developed.

A process which consists of selecting individual high promise projects, in itself a difficult enough exercise, might at first sight be thought to lead to the most effective use of R & D resources. But projects are truly independent only when unlimited resources are available. In practice, however, funds are restricted and projects are competing with each other for laboratory equipment, technologists, materials and, perhaps the scarcest resource of all, good project management. The aim of maximizing the contribution from the whole R & D portfolio may, therefore, occasionally result in the rejection of a project which might otherwise appear attractive because its high consumption of a particular resource leads to the starvation of other projects also requiring that same resource.

As with a portfolio of financial investments, a portfolio of projects develops certain characteristics, either as an act of policy or by chance. If compiled with the sole purpose of maximizing the expected financial return on the project investments and minimizing their mutual interference the portfolio may exhibit no other factor binding it together. Alternatively, it may develop a strong orientation towards a particular technology, product line or market which emerges without any serious consideration of its implications for the future of the business. This can be likened to an opportunistic policy which supports only those projects likely to provide the maximum short-term profit. Such a policy can succeed, although on the other hand it can easily lead to disaster in the long run through neglecting to lay the foundations for the products needed to meet major changes in the market environment. A solution is needed in which a portfolio is deliberately constructed in such a way that it supports the corporate objectives while at the same time permitting sufficient flexibility to accept outstanding opportunities when they occur.

A decision-making process which integrates R & D decisions with the corporate strategy is illustrated in Fig. 2.3. The top of the diagram shows the combination of environmental forecasts and capability analysis in the formulation of the corporate strategy; the bottom shows the inputs to the project selection process. The two parts are linked together through the R & D strategy. It should not be thought, however, that this is a rigid 'top-down' process, for this would have similar weaknesses to a 'bottom-up' approach where the selection of projects determines the future character of the company. To be meaningful it must become the vehicle for a continuous dialogue between top management and R & D management.

The R & D strategy thus performs for R & D a similar role to that of corporate strategy for the company as a whole (Table 2.1).

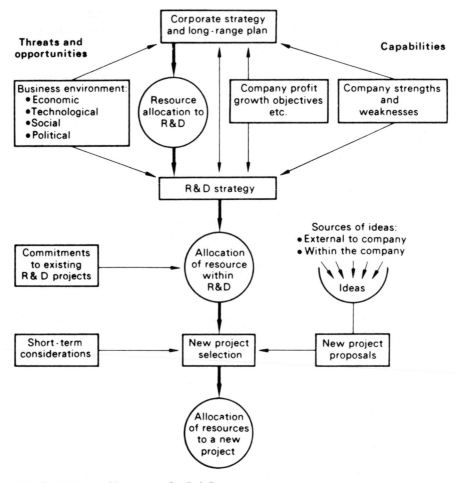

2.3 The decision-making process for R & D

Table 2.1 *Comparison of roles of corporate and R & D strategies*

Area of influence	Corporate strategy	R & D strategy
Resources	Allocation between functions (marketing, production, R & D, etc.)	Allocation between projects
Objectives	Related to business environment	Related to corporate environment
Business areas	Product/market strategy Product/market mix	Technology/product strategy Portfolio balance
Timescale	Balance between long/medium/short term	Balance between long/medium/short term

Resource allocation within R & D

'Gap analysis'

The corporate long-range plan will specify profit objectives for a number of years ahead (Fig. 2.4). This curve represents the summation of individual contributions from a population of products which is ever-changing in their

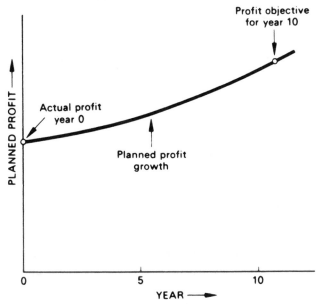

2.4 Corporate profit projections

number, nature and relative importance. Since products have a limited life a projection of the anticipated profit contribution from products existing at the beginning of the period will fall away. The shortfall between what can be expected from these products and the requirements of the corporate plan (the shaded portion in Fig. 2.5) must be filled by new products. The role of R & D is to ensure that the new products are available when required and are of a kind which will make an adequate profit. 'Gap analysis', the name used by corporate planners to describe this process, not only shows the magnitude of the task confronting R & D but also indicates a more rational basis for estimating the funds required for R & D than approximations based on past or present performance or competitors expenditure. Although gap analysis is attractive conceptually problems often arise through top management's reluctance to accept that a gap exists. A major factor which often clouds the issue is the superimposition of the economic cycle upon the long-term trend [5].

The aggregation of the expected profit contributions of new R & D projects cannot, however, be matched against the requirements revealed by the gap analysis. This is because of the uncertainties of the estimates and the expectation

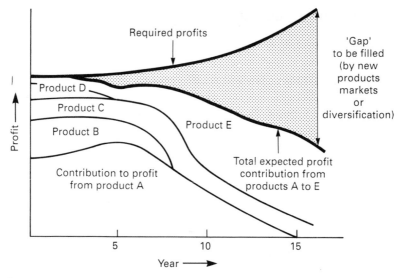

2.5 Gap analysis — showing shortfall between required profit and the contribution to be expected from existing products and markets

that only a small proportion of projects will result in commercially successful products. The best that can be hoped for is that the magnitude of the R & D effort is made commensurate with the demands placed upon it and that the funds are allocated in a way planned to give a high probability of yielding suitable new products at the time they are needed. This adds a new dimension to the balance required in the portfolio — time. In spite of the difficulty in predicting project outcomes, an effort must be made to avoid undue bunching or gaps in the completion of product developments.

Product life-cycles

Most products exhibit a similar pattern when their volume and profit histories are plotted (Fig. 2.6). Initial growth in sales volume after launch is slow. The product is not widely known and a high price makes it attractive to only a small number of customers who are prepared to buy the unproven product for one of several reasons. Perhaps it satisfies a particular need, they like to possess something different from their neighbours or they are innovators who like to try something new or novel. The high price reflects the costs of development, marketing launch, and production inexperience as well as the matching of demand with a production volume likely to be low until manufacturing output builds up to the planned plant capacity. Although this is the usual situation, there are exceptions where a minimum plant size for economical operation (e.g., process industries) creates the need to increase volume quickly; this may remove the opportunity to set a high initial price and may consequently delay the achievement of financial break-even.

The behaviour of the profit curve reflects both the launching costs and the high initial margins. After a period of unprofitable trading due to the low volume, the profit curve rises more sharply than the volume curve. The shape of this curve is unaffected by the addition of the R & D to the launching costs, although the period to the recovery of all costs would be longer.

Market growth attracts competition as other companies develop similar products. This causes a steady erosion of the price which increasingly offsets the manufacturing cost reductions resulting from the 'learning curve' effect. Profit margins shrink although the increasing sales volume ensures the continued growth of total profit, although at a reducing rate.

Eventually the sales volume falls as competitors introduce new models of higher performance or at lower prices. In order to remain competitive the company must launch its own new product at or around the time of maximum profit. Timing is important because of the long lead times; thus development programmes must be carefully synchronized with the expected product life cycle of the existing model. Premature launch is wasteful in that it does not fully exploit the profit potential of the current product. Late launch incurs short-term loss of profit and may result in a long-term loss of market share which may be difficult to recover.

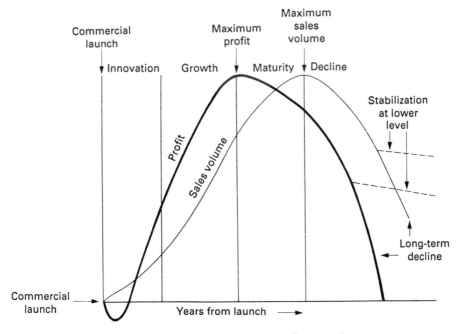

Notes: 1. Profit and sales volume life cycle drawn to different scales.
2. Profit life cycle rises and falls ahead of volume life cycle because of:
   • High margins during innovation phase.
   • Increasing competition during mature phase.

2.6 Profit and sales volume – life-cycles for a typical product

Study of the product life-cycles indicates a number of considerations to be borne in mind when formulating a strategy for R & D:

1. The different behaviour of the profit and volume life-cycles.
2. The curves show the performance of successful products. While the high initial profit growth following breakeven is attractive, the risk of failing to achieve market acceptance is also high. The breakeven may in fact never be reached.
3. Initial success relies heavily on the ability to develop innovative products, whereas continued profitability rests more heavily upon marketing and manufacturing expertise to maintain sales volume with low production costs.

Single product companies

Although gap analysis is necessary for understanding the need in a multi-product firm, discussion of the simplified example of a single product company brings into relief the problems to be considered in formulating a strategy. Figure 2.7 gives a typical example of the situation facing a company manufacturing a single product, the nature of which goes through a succession of changes sufficiently important for each to be regarded as a separate product in its own right.

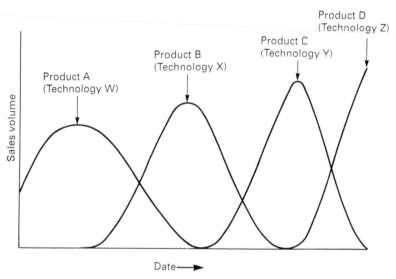

2.7 Product lives – technological substitutions

In Fig. 2.7 it will be noted that the product lives are shortening, a common feature when technology is advancing rapidly. When we explore technology life-cycles in Chapter 8 it will be seen that as a particular technology matures the life-cycles of products based upon it are likely to lengthen. Thus it is not possible to state that product life-cycles are shortening as a generalization; each situation must be examined in detail. A lengthening of product life-cycles will occur only, however, in the absence of a new technology which substitutes for the old. The situation in audio recording exhibits a pattern similar to that in Fig. 2.7. After the original recordings of Edison made on a cylinder, the development of the flat gramophone record remained essentially unchanged for over a century. During this period there were a number of incremental innovations. Bakelite was replaced by vinyl. The steel stylus was replaced by fibre and later by diamond. EP and LP discs were introduced. But after this long period of relative stability the rate of change has accelerated during the past 40 years. First, the tape substituted for a large part of the vinyl disc market. Then at reduced intervals the compact disc (CD) was introduced, followed rapidly by the development of digital audio tape (DAT), and Sony's mini-disc in the early 1990s; the commercial success of the latter two developments are unproven at the time of writing. It should be noted that the innovations of the 1980s and 1990s were based upon the application of technologies originally developed for different purposes.

The replacement of an established product may cover a considerable period of time, with alternative products competing for market share. In 1991 vinyl discs, although losing market share rapidly, were still being sold in competition with both tapes and CDs. Thus the appearance of a new product does not necessarily mean the immediate loss of the non-innovating firm's market. There may be plenty of time for the development of a competitive product, perhaps possessing advantages over the original innovation, and to choose the best time to enter the market. This should not be allowed to breed a complacent attitude for the rate of substitution is not always slow, as we have seen with the example of the float glass process. A judgment must be made on the rate of substitution and this will have a major influence on the R & D policy to be adopted. Rapid replacement is closely related to the advantages the innovation gives the customer either in added performance or cost savings. Comparison of the rapid rate at which the transistor replaced the thermionic valve with the much slower substitution of conventional by nuclear power stations shows clearly the influence of these factors.

Summarizing, it can be seen that it is important to:

1. Anticipate the introduction of new technologies which may threaten established products.
2. Estimate the date at which the new products are likely to appear.
3. Estimate the rate at which the new product will capture the market.
4. Decide when to enter the market with a competitive new product.

Timing is an essential consideration for what is a critical strategic decision. Technology forecasting (see Chapter 8) can make a valuable contribution to these vital strategic decisions.

The requirements for profit growth can now be combined with the contributions obtainable from each of a family of successive products (Fig. 2.8).

But the shapes of the individual profit curves are not preordained, they result from a series of R & D decisions. They are made to happen, and the shapes of the curves are determined by the allocation of resources to individual R & D projects. The arrows in Fig. 2.8 show how this is brought about:

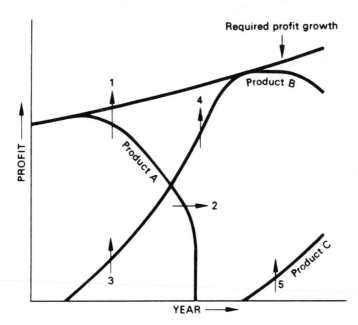

EFFECT OF R&D INVESTMENT

1. Short-term development of existing product
2. Extension of product life
3. First into market with new product
4. Late entry with new product
5. Long-term development of third generation product

2.8 Profit growth from successive products – one-product company

1. Short-term development of the existing product

The current product may remain on sale for some considerable time. Left in its present form without further development, profitability can be expected to fall under the impact of the competitive pressure on price and improvements introduced by other manufacturers. This threat can be met by devoting R & D effort to:

1. Product improvement aimed at capturing a larger market share through detailed refinements using the company's traditional technology or by the application of new technology, materials or processes.
2. Improvement in profit margin through process developments which reduce manufacturing costs.

2. Medium-term development of the existing product

It may still be possible to extend the useful life of the product by additional R & D. Frequently, major improvements in performance have been introduced which were stimulated by the threat of a replacement.

If this is possible, it may be a highly desirable policy to adopt. The manufacturing facilities for the old product usually represent a considerable capital investment. Irrespective of the book value of these facilities, which depends upon the company's depreciation policy, their worth in real terms is likely to be negligible if they have to be scrapped completely when production is discontinued. They represent what the accountant terms a 'sunk' cost. Further product development to retard the rate at which the market declines extends the useful life of this equipment. In this situation the 'true' profit to the company is represented by the difference between the sales value and the marginal cost of production less the additional R & D investment required to extend the product life. The depreciation of the capital equipment is neglected since it would otherwise have been scrapped. The accountant may still compute a book profit incorporating an allocated allowance for depreciation but this figure is irrelevant in the decision-making process.

When calculated in this way, the returns obtainable from additional R & D to extend product life can be considerable. Even when the product continues to decline this investment may result in an increased share of the reduced market.

The financial attractions of continued investment in a declining product must not, of course, be allowed to hide the obvious dangers. It must be undertaken as part of a deliberate policy which ensures that adequate provision is made for the development of the product which will eventually replace it. If this is neglected the company's competitive position can quickly erode in the face of a new technology with a rapid diffusion rate. Slow reaction to the introduction of electronic watches caused a rapid decline in the size and profitability of the Swiss industry, for long the world leader.

3. Early introduction of a new product

R & D effort and expertise can be focused upon the achievement of technological leadership, and the introduction of innovative new products before competitors. This is often called an 'offensive strategy'. The Glaxo anti-ulcer drug Zantac, the Sony Walkman and float glass illustrate the rewards awaiting such innovations when successful. But the risks of technological failure or bad timing cannot be ignored. These dangers of innovation are borne out by the experiences of ICI with Pruteen, Rolls-Royce with Hyfil compressor blades, Thalidomide, the video-telephone, and early investments in cable and satellite television.

This is essentially a high risk strategy. It also offers the opportunity for a high financial payoff.

4. Late introduction of a new product

The extensive substitution period for many products offers the possibility of deliberately refraining from entry into a new market until it has been pioneered by a competitor. This is termed a 'defensive strategy'. In return for the security given by the knowledge that a market for the product exists, the initial high rewards are sacrificed. This is thus likely to be a relatively low risk/low payoff strategy.

Because it is less demanding to be the follower than the innovator the R & D investment is likely to be correspondingly lower both in respect of the cost of developing the new product and the involvement of high calibre technologists.

5. Long-term projects

Further into the future the second-generation product itself will become obsolete and need to be replaced. In rapid growth technologies product lives can be relatively short, measured in years rather than decades. This represents a number of R & D investment choices, particularly if it is not clear what form the longer term product will take; these range from the initiation of a major R & D programme to 'wait and see'. The alternative actions which can be taken include:

1. 'Technology forecasting' and 'technological monitoring', to ensure that significant developments are identified and interpreted.
2. A minimum 'foot in the door' investment to build up expertise and R & D personnel in any new technologies involved. This should enable a rapid response to be made at some later date when new developments appear which look like posing a major threat.
3. A major R & D programme, maybe at the expense of medium-term projects, in order to seize the initiative in what might be a significant innovation.

One of the most critical choices in formulating a technological strategy is that between the development of entirely new products and the incremental improvement of the current products. The new product decision is the more risky in the short term since it involves moving into uncharted territory by applying technology new to the company. Nevertheless, no product is immortal and a time eventually arrives when it becomes impossible to sustain its competitiveness. But incremental improvement does have advantages. It is low risk and builds upon past investment, known technology and experience of the market. But many companies which have relied too heavily upon incremental development have seen their markets eroded and have found it difficult to regain their market share when a new product is eventually launched. Conversely other companies have introduced a steady stream of innovative products without fully exploiting their development or profit potential. The correct timing of the transition is of the essence of a good technological strategy.

In an ideal world where unlimited resources were available it would be possible to support sufficient projects to enable the exploitation of all five of the alternatives outlined above. In practice, this is rarely the case and a choice must be made, a choice which means the sacrifice of one opportunity in favour of another. Furthermore, strategic considerations might sometimes dictate preference being given to a project which appears less favourable than another in terms of short-term profitability. The development of an existing product may, for example, leave insufficient resources for the conduct of an active offensive strategy or a choice may have to be made between the development of new products for the medium or the long term. Generally, sufficient resources will be available for work in several areas. But it is necessary to develop guidelines for deciding what will be neglected and the relative priorities to be given to the areas in which work is to be done.

Analysis of resource allocations

The reasoning applied to the analysis of the single product company is equally applicable to a multi-product concern. In this case, however, the multiplicity of possibilities makes it difficult to identify them as specific project alternatives. It will also be necessary to take into account the relative importance to be attached to the different product areas.

This difficulty can be overcome by analysing the spending opportunities under general headings. The selection of the appropriate breakdown for expenditure will depend upon the nature of the company but is likely to include the division between short-, medium- and long-term work; between existing and new products; and between existing and new technologies (Fig. 2.9). Other possible divisions might be between product and process development or between product areas. Figure 2.9 can be likened to a cash box representing the R & D budget, the smaller cubes being ways in which the budget can be allocated. The portfolio balance and the long-term contribution for the R & D department will depend upon the appropriations made to the smaller cubes. Only then need attention be turned to the selection of individual projects to meet these requirements.

It is a useful exercise to examine the previous year's expenditure in this way. If this has not been attempted before some difficulty will be experienced in deciding how past project expenditures relate to the headings. This need not be a major obstacle; great accuracy is not essential since the main purpose of the exercise is to find out in general terms what the current situation is. Even so the results may reveal some surprising information which had not been fully appreciated previously. It might be found, for example, contrary to all expectations, that very little effort was devoted to developing new technologies. Or, in a company where the total effort is divided between a central research laboratory responsible for long-term R & D and divisional laboratories concerned with product development, it may become apparent that the medium term has received relatively little attention.

Having established how resources are currently being allocated the question arises of what to do with the result. Should the emphasis be changed? And if so, in what respect? It is the role of an R & D strategy to indicate the answers.

One reason why an analysis of this type can be of great value is that it may take a considerable time to accomplish a change of emphasis. Because in most laboratories many projects continue through several budget periods a high

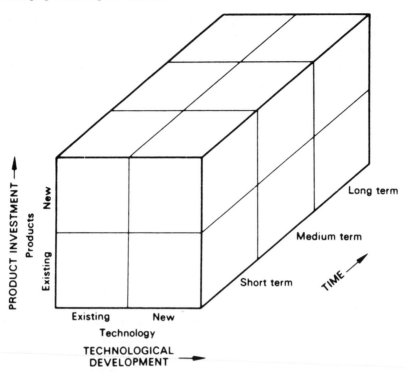

2.9 Categorization of the R & D budget

proportion of the funds available in the current year will already have been earmarked for ongoing projects. Thus the amount remaining for new projects, which can change the balance of the portfolio, is likely to be limited. This restricted freedom to manoeuvre means that it is not possible to respond speedily to changes which have been identified as desirable without causing severe dislocation to the existing programmes.

The analysis undertaken so far has done two things:

1. Identified alternative ways in which it is possible to allocate R & D resources.
2. Established how this is being done at the present time.

If a deliberate strategy has not been followed in the past it may still be possible to reveal an implicit strategy in the pattern of project decisions that have been taken. On the other hand it may be impossible to detect any connection between an apparently random collection of projects. The next logical step in this process is to select what appears to be the most appropriate strategy for the future and to use this as a basis for the selection of projects to bring about the changes in emphasis deemed necessary.

Factors to be considered in the formulation of an R & D strategy

Referring to Fig. 2.3, it can be seen that there are three main informational inputs to the R & D strategy – environmental forecasts, capability analysis and the corporate strategy. Working systematically down the diagram, project selection appears to be derived from the R & D strategy. But we have said previously that strategy formulation is an iterative process and it would be wholly unrealistic to take no account of the potential projects which are available when considering strategy. For it is the projects which make the achievement of the strategy possible. Nor is it possible to ignore the likely allocation of resources to R & D since the various strategies will involve different levels of expenditure. If the allocation is inadequate some strategies, however desirable they might otherwise appear, will have to be excluded.

The major factors which need to be considered in the formulation of an R & D strategy will now be examined.

Environmental forecasts

The need for environmental forecasting in relation to corporate planning has already been discussed. The main purpose was seen to be the establishment of what might be done to exploit the opportunities or meet the threats arising from possible future changes in the business environment. The R & D strategy which in many ways can be regarded as an extension and integral part of the planning process uses forecasts in a similar fashion.

R & D will be primarily concerned with the changes in technology which will occur in the future. However, it will be shown that it is not possible to forecast technology in isolation from a host of other factors, some economic, some social and some political. Since all these trends interact in a highly complex fashion technology forecasting must cover much of the same ground as the general business forecasting activity.

The desirability of following any particular line of policy is influenced by the amount of competition which is likely to be experienced. Knowledge of how competitors will respond to environmental change would be extremely valuable. This knowledge one is never likely to have, but nevertheless it is possible to deduce what logical reactions might be expected from them by matching an analysis of their capabilities with the forecasts. Although there is no guarantee that either their forecasts or logic are the same as one's own an insight into their possible competitive strategies may be helpful in avoiding undesirable decisions.

It is easy to assume that there is no doubt regarding the identity of the competition. But frequently the most dangerous competition appears from an unexpected quarter. A valuable outcome of the forecasting process may well be the revelation of new competitive technologies and businesses arising from them. This can be illustrated by taking the example of a company which manufactures metal cans. The top management may regard their business in one of the several ways defined by the writers of corporate strategy, either in product terms (metal cans), in generic terms related to function (packaging), or in terms of the technology employed (metal forming). Although it might be thought wise to keep all definitions in mind it is rarely done. In these circumstances it might seem obvious to think of the company as a manufacturer of metal cans and the competition to be from other manufacturers of metal cans. This line of thought would lead to a concentration of R & D thinking solely on the advances in metal canning technology or metal forming. Consequently, the competitive threat from other materials, such as lightweight non-returnable glass bottles, might be overlooked until they have reached a late stage of development or make their market debut, by which time it might be too late to take avoiding action.

In this way we observe the not uncommon occurrence of 'innovation by invasion'. It was not, for example, an established manufacturer of bedsheets in the cotton or linen industries who introduced nylon bedsheets, but a newcomer. Similarly a number of new businesses based their companies on plastic kitchenware at the expense of the traditional manufacturers who continued to make their products from metal. Nowadays microelectronics represents a threat to almost any company in the electromechanical industry. Thus in analysing threats and opportunities attention must be paid to the strategic threats posed to one's own product by the application of technologies not traditional in the industry; also one should examine how one's own technology might be used offensively to invade another industry.

It is not only technological industries which need to remain on the lookout for threats from new technology. Technological growth has historically led to the replacement of manual labour and natural products. Thus companies which are labour intensive or dependent upon the sale or processing of natural products must also constantly monitor these trends.

The value of environmental forecasting in strategy formulation can, therefore, be seen in three main areas:

1. Identification of future threats and opportunities.
2. Avoidance of technological surprises.
3. Identification of new competitive technologies and businesses.

Comparative technological cost effectiveness

In Chapter 1 reference was made to the suggestion that technologies as well as products have life-cycles. The body of knowledge in a technology increases until it reaches a point where further research yields new knowledge whose incremental commercial benefit becomes negligible. When this stage is reached investment in a new branch of technology is likely to offer far more promising opportunities for new products, processes or product improvements. This indicates a shift of resources into the cubes labelled 'New technology' in Fig. 2.9. An example of the type of reorientation that may have to occur can be drawn from the electrical industry where it is becoming increasingly difficult to obtain significant performance improvements from conventional electrical technology. However, the commercial benefits which might be derived from superconductivity appear of a different order but require the technology of cryogenics and the employment of low temperature physicists rather than electrical engineers.

Such a reorientation is not easily achieved. For instance, the relevance of advances in a new technology may be missed completely if the company has no experience in it. This can be particularly true of a business unused to change or where the scale of R & D and the spread of disciplines is small. 'Technology forecasting' can help in spotting the relevant trends, but it still leaves the problem of what to do about them unanswered. The slow response rate to change by an R & D department results from the nature of its major asset, people, who have a particular discipline or specialization. Their skills and knowledge have been acquired over a long period and most of them find it possible to change these only within narrow limits. Nor can their number normally be changed quickly for, except in periods of severe economic pressure, it is unusual for technologists to be made redundant even when their discipline is becoming of decreasing value, a fact which itself might have escaped notice. Thus they are likely to remain in the R & D department and continue to do work absorbing resources which could be deployed more profitably elsewhere; for opportunities for further R & D will always remain and new projects will be proposed. The potential for economic developments from the technology will become exhausted before the supply of projects for the extension of knowledge in it.

These difficulties are compounded by the likelihood that they will occur in the large mature organization which is no longer growing rapidly; growth can accommodate change more readily. The acquisition of additional resources enables the imbalance between technologies to be redressed by the introduction of new specialists; but the mature organization, aiming to provide its employees with security of employment, has little freedom to manoeuvre.

Certain actions can, however, be taken to alleviate this situation:

1. *Retraining.* It is slowly being recognized that in a rapidly changing world, education does not cease on graduation. Nevertheless, major re-education is difficult to achieve unless planned on a long-term career basis.
2. *Re-employment.* Many R & D personnel can still make a valuable contribution in other functions (e.g. manufacturing and marketing) even when their own specialist knowledge has become redundant.
3. *Recruitment.* The balance between technologies can be modified by the recruitment policy provided it aims to satisfy the needs of the future rather than short-term pressures. This is another area where forecasting can be useful in anticipating future changes which will require to be reflected in the staffing of the department.

Risk *v* payoff

The degree of risk inherent in different R & D strategies has already been noted. The corporate strategy also is influenced by top management's attitude to risk, although it may not be stated explicitly. Usually it can be inferred since the strategy reflects their system of values. There are two aspects of risk to be considered in R & D arising from its examination in relation to the portfolio and to individual projects.

The risk inherent in the portfolio must be a major concern of the R & D strategy and should reflect the corporate attitude. But this risk is then spread over a mix of projects each of which is associated with its own level of risk. In the multi-project laboratory both offensive and defensive strategies, with their different levels of risk, can be followed, the proportion of each being adjusted in accordance with the risk propensity acceptable to the corporate and R & D strategies.

Hertz [6] and others have developed techniques of risk analysis. These enable a rational mathematical examination of the expected payoffs from alternative projects to be weighed against the subjective probabilities of the occurrence of the outcomes. This approach is illustrated in Fig. 2.10 which shows the patterns to be expected. Project A, high risk, exhibits a flat profile having a high expected value for the payoff but also the possibility of a very high profit or even a loss; on the other hand Project B, low risk, has a much sharper profile with a low variability, but the expected profit is also lower.

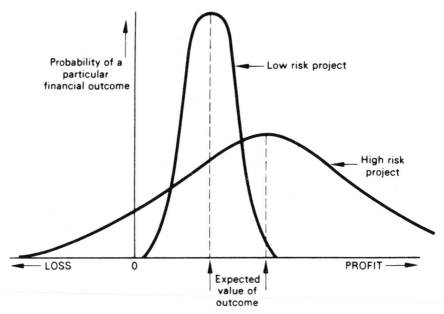

2.10 Risk profiles for two projects

'Risk analysis' might lead one to deduce that the large company, able to spread its risk over a number of projects, would favour an offensive strategy, whereas the small or one-product company would favour a defensive strategy because of its lower risk, particularly in circumstances where no possibility of loss can be entertained. However, there are grounds for believing that this logic does not always apply in practice and that the reverse effect occurs. This may be because the top managements of the large mature organizations are more risk averse than their smaller entrepreneurial competitors. It may also result from their inability to create a risk-taking environment for their operational managers, not because these managers are naturally disinclined to take risks but because they believe that the personal penalties for a failure far outweigh the rewards from success. Thus there is an attenuation of the willingness to accept risk as one descends the hierarchy of the organization, increasing with the distance between top management and the decision-taker. Occasionally, however, the large company is faced with a situation similar to that of the single product company. Rolls-Royce, with the RB 211 project, was in this position. In judging this company in the light of their subsequent failure it is appropriate to ask whether there were satisfactory alternatives and whether the risks inherent in the decision were properly analysed. It may well be that the strategy followed was the correct one and that the risks were accepted knowingly. Business cannot be divorced from risk.

Risk is a consequence of uncertainty which, in technological innovation, arises when changes are made in the technology, the products, the manufacturing processes or the markets. This is minimized when innovation occurs in only one of these factors at a time, for example by applying a new technology in an existing product or designing a new product for an established market using current technology; it is greatest when a new technology is embodied in a new product for a market new to the company using a new production process. The possible alternatives are shown in Table 2.2. These considerations do not

Table 2.2 *Risk related to the degree of change*

Change in	Low	Medium	High	Very high
		Risk		
Technology	× – – –	× × × – – –	× × × –	×
Product	– × – –	× – – × × –	× × – ×	×
Market	– – × –	– × – × – ×	× – × ×	×
Production process	– – – ×	– – × – × ×	– × × ×	×

Key: X New   – Current

necessarily imply that medium- or high-risk projects should be avoided since, if successful, they can yield the greatest benefits. However, the alternatives should be analysed critically. Most managements would favour a cautious strategy wherever feasible, where only one, or at most two, of the variables are changed at one time.

Planning can never remove the risk from business decisions. But at least one can hope that a process of rational analysis will enable the avoidance of most of the obvious pitfalls and an assessment of the risks inherent in the identifiable uncertainties. However careful the analysis, there is always likely to be something which is either overlooked or could not have been anticipated.

Capability analysis

Before deciding upon a strategy it is essential to make a realistic appraisal of one's own strengths and weaknesses. Wishful thinking must play no part in this exercise. Although it is useful to analyse past and present capabilities, these may not be relevant to the needs of a future which may demand significantly different attributes. Today's strengths may have little relevance to the future unless useful outlets can be found for them. Nothing is impossible to change and it may be argued that the weaknesses can be remedied. This must certainly be attempted but it is likely to take a long time and turning a blind eye to them can lead into a fatal trap.

The relative strengths of the functions within a business have an important bearing on the formulation not only of the corporate strategy but also of the R & D strategy. For example, a company whose main strengths relative to the competition lie in marketing and production rather than in technology is unlikely to be successful in new product innovation. Conversely, R & D strength would suggest an offensive strategy where a stream of innovative products makes up for possible shortcomings in marketing and high manufacturing costs which are less important in this situation.

The R & D department, however, possesses its own unique mixture of strengths and weaknesses which should be analysed. The identification of a technological opportunity or a creative idea does not necessarily mean that the ability exists to carry them through to successful commercial exploitation. The analysis can take the form shown in Fig. 2.11. The first column gives examples of the types of factor to be considered; the factors themselves will differ to some extent from one company to another according to the characteristics of the industry. The column 'Audit of present position' contains an assessment of the current position represented in the form of a profile to aid in the identification of areas of particular strength or weakness. The third column 'Technological capital for the future' is more difficult to complete. Its purpose is to evaluate the company's technological capabilities in relation to the demands upon it to meet its future objectives. This presupposes that these demands are known, but they are derived from a still undetermined strategy or project portfolio; the process of strategy formulation and the assessment of technological capital must consequently form part of a combined exercise. The concept of technological capital focuses the mind on the future and the real worth of the resources available. In this sense a laboratory full of expensive equipment or highly qualified personnel may represent no technological capital unless they can be invested to yield a future benefit, indeed they may become a liability rather than assets if they cannot be disposed of. The difference between the audit and the technological capital

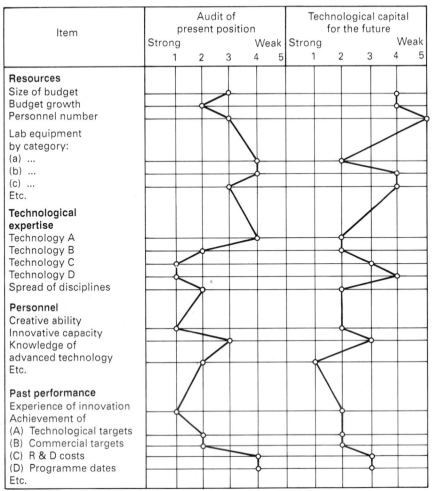

Note: 'Present audit' represents an evaluation of current capability.
'Technological capital' can only be related to a stated objective for the
future and represents the anticipated capability of meeting that objective

2.11 Technological audit

profiles can now be seen to give an indication of the changes required to satisfy
the chosen strategy.

In analysing the technological resources it will often be found that one or
two items will distinguish themselves as outstandingly strong or weak. These
may hold the key to ultimate success or failure. Ward [7] defines the terms
'differentiated assets' and 'differentiated liabilities' to describe these attributes:

In defining corporate identity, it is useful in the first instance to identify those
assets which in combinaton distinguish an organization from every other. It
is these assets that provide a business with any competitive advantage it may

have. The market is common to everybody; a company's potential is unique. At the same time, it is necessary to examine the company's 'differentiated liabilities', since they may represent a limiting constraint.

A particular technological strength or an individual person may be a company's differentiated asset. The competitive strength this gives to a business may be of outstanding value. The Polaroid company, for example, has been founded and sustained by the inventive genius of Dr Land. When Kodak decided to market its own self-developing film for use in Polaroid cameras, the multimillion dollar project was later abandoned. The lesson is clear. Exploit your own differentiated asset to the full, but think carefully before challenging a competitor's, however strong you are.

Technological custodianship

The raw material for all technological innovation is knowledge. This may be knowledge required to solve current problems or that needed for incorporation in new projects to further the corporate strategy. In addition, new technologies or developments in existing technologies may in the longer term provide opportunities for entirely new products or suggest areas for business diversification. This 'strategic push' can come only from the knowledge base residing in R & D.

The establishment and maintenance of the knowledge base is an important corporate responsibility which only the technologists can fulfil. To a great extent it relates to technologies or their advances which meet no current clearly defined need but which may have a role to play in the future. An analogy can be drawn with the finance director who is responsible for ensuring that financial resources exist over and above current budgetary requirements in order to be able to exploit unspecified future opportunities as they occur. In a similar way R & D is the custodian of the company's technological resource. This responsibility cannot be fully discharged unless some funds are allocated to the acquisition of new knowledge, on occasions supported by a limited amount of applied research, in order to evaluate its potential. This custodial function must be a personal role of the R & D director for only his judgment, formed in consultation with colleagues, can assess the potential and the need to build up knowledge in an emerging technology.

The technology portfolio

The knowledge base required by a company is expanding rapidly in both depth and breadth. Most firms now find that they must acquire an expertise in technologies not previously incorporated in their products. A company's technology base can be categorized as:

1. Core technologies.
2. Complementary technologies.

3. Peripheral technologies.
4. Emerging technologies.

The *core technology* is central to all or most of the company's products, expertise in which was the main contributor to past success. The culture of the laboratory is heavily orientated to that technology which is likely to be the discipline of the R & D director and many of his senior managers. It dominates the corporate thinking.

A *complementary technology* is an additional technology, knowledge of which is essential in product design. In many industries today materials science may be the main contributor to future success, for example in semiconductors or the ceramic engine. In other industries hybrid technologies such as opto-electronics, mechatronics and the use of platinum in pharmaceuticals are becoming of increasing importance. This raises some serious managerial issues since it is often difficult to recruit or retain high calibre personnel in the complementary technology. They may be made to feel that they are in an alien culture and their career prospects are limited. This manifests itself in such cases as the poor mechanical design often evident in electronic products or electronic failures in automotive products.

A *peripheral technology* is defined as one not necessarily incorporated in the product but the application of which contributes to the effectiveness of the business. The application of IT in managerial systems and new process technology in manufacture would fall into this category.

Emerging technologies are those that are new to the company but may have a long-term strategic significance. While not of immediate importance in product development it may be necessary to build up a competence or undertake exploratory research in order to provide the strategic flexibility to respond quickly when an opportunity arises. One example is the current investment by Japanese electronic companies in biotechnology.

A framework for analysis is shown in Table 2.3. It directs attention to the future and indicates where a change in emphasis may be required. This provides a basis for consideration of the likely timescales, the allocation of resources and how the transition should be managed. A similar table can be drawn showing the current expenditure in each category and used as a basis for examining how these should change in the future.

The technology portfolio analysis directs attention to the needs of the future. This must be combined with an evaluation of the competence of the organization to exploit those needs. Durand [8] proposes a systematic approach using a programmes/competencies matrix in order to construct an inventory of expertise 'which could improve the efficiency of R & D resource usage, identify the laboratory's strengths and weaknesses, direct R & D into hitherto neglected channels, assist individuals to identify and evaluate their own expertise and justify obtaining funding for building expertise in shortage areas'.

Table 2.3   *Technology portfolio analysis*

Technology	Relative importance					
	Current			Future		
	High	Medium	Low	High	Medium	Low
Core						
A	×			×		
B	×				×	
C		×				×
Complementary						
D		×		×		
E			×	×		
F			×		×	
Peripheral						
G			×	×		
H			×		×	
Emerging						
J			×	×		
K			×		×	

Selecting the R & D strategy

A great deal has been written and spoken about the formulation and selection of a strategy but in practice it is not often that a clearly defined decision point is reached. It is the analytical process that is important and leads gradually to the emergence of a strategy. This is particularly true in respect of an R & D strategy which results from a continuous process of evaluation and analysis of the various interrelationships which link together the corporate strategy, environmental analyses, capability audits, the portfolio and individual projects. Nor is one usually faced with an exclusive selection between strategies, but rather a decision of where to place the emphasis. For example, the choice will not be between an 'offensive strategy' or a 'defensive strategy' but the proportions of the available resources to be allocated to each of them.

We shall now pull together the earlier discussion and indicate how the analyses will influence the choice of emphasis between strategies.

Offensive strategy

This high risk, high potential payoff strategy demands considerable skill in technological innovation, the ability to see new market opportunities in technological terms and the competence to translate them swiftly into commercial products. In most cases a strong research orientation coupled with the application of new technology will be necessary.

A number of studies have been carried out on the effect of company size on innovation (e.g. Jewkes, Sawers and Stillerman [9] and Cooper [10]). They suggest, that in a number of industries, even when they are dominated by a few major manufacturers, it is still possible for the small company to adopt a successful offensive strategy. Many of the major innovations of the last two decades have been made by small companies. The factors discussed in Chapter 1 tend to occur naturally in the small company, particularly during a period of entrepreneurial growth where the top management is likely to be closely attuned to market needs and the chief executive may personally be the project champion. Consequently, the organizational climate in the small company tends to be favourable towards innovation as are the informal management style and the short internal lines of communication.

The small firm's inability to support a large R & D department might suggest that it would not be able to provide the resources necessary for an offensive strategy. Although true of some technologies, there are many others where the small company can concentrate on one project an effort comparable to its large competitor whose resources are spread over a number of projects. But the large concern is primarily interested in products which can make a worthwhile contribution to its total profits. It is, therefore, unlikely to look kindly upon new products which may take several years to build up a sufficient sales volume to contribute a significant profit. This same profit, however, in the hands of the small company makes a relatively much more important contribution to overall profitability.

Some large companies may have little alternative to adopting an offensive strategy. The market leader in an industry dominated by a few major companies is highly vulnerable, since his position can be destroyed by the introduction of a technologically superior product by a competitor. Although it is possible to rely upon superior marketing or manufacturing expertise as the base for a strong counter-attack, this policy may not yield results before market leadership has been severely eroded. In general, it may be concluded that there is a strong argument for such businesses to follow an offensive strategy.

Between the extremes of the small company and the market leader there lies a large number of companies where no compelling reason exists for favouring an offensive strategy. There is a freedom of choice which can be influenced by managerial judgment based on a careful analysis.

Defensive strategy

Essentially low risk low payoff, a defensive strategy is suitable for a company able to earn profits under conditions of competition. To do so, it is necessary to capture a market share and sustain profit margins through low manufacturing costs when there is pressure on prices. Thus a defensive strategy is likely to commend itself to a company whose business strengths are in marketing and production rather than in R & D. At the same time it must still retain sufficient technological effort to enable a swift response to innovation by a competitor.

In general, the laboratory strengths are likely to lie more towards the development end of the R & D spectrum, since new technology usually is less important than the ability to exploit an existing technology to the fullest extent.

Licensing

Licensing, sometimes called an 'absorbtive strategy' presents many opportunities for commercial gain by buying the fruits of another company's R & D investment. Technological innovation and internal investment in R & D are not synonymous and there is little gain from rediscovering what can be obtained from another source more cheaply. Even the largest companies cannot afford to examine all the technological alternatives or produce all the worthwhile innovations. Nevertheless, many companies are markedly reluctant to pursue an active licensing policy.

Experience shows how important it is that companies, even the R & D leaders, should not restrict their commercial opportunities by relying upon a strategy based solely on innovations generated by in-house R & D. This is not to underrate the value of R & D, for internal technological strength makes an important contribution in identifying what should be licensed. Licensing, then, is a supportive strategy. Limited financial resources may present a small company from fully exploiting a major innovation. Licensing permits the maximum benefit to be gained without placing an undue strain on the company's finances. Sulzer Brothers for example, has always followed an active licensing policy for this reason and has over thirty licensees for its marine diesels, world-wide.

Licensing of one's own major innovations is also valuable for a company following an offensive strategy. The prospect of capturing a whole market by putting the competition out of business is very slender, although IBM came very close to doing so for many years. Faced with a major innovation competitors will react by investing heavily in their own offensive research or by attempting to navigate around the patent protection. However, the offer of a manufacturing licence for the innovation is likely to divert effort from their own research, thereby largely disarming them. Pilkington Ltd followed this policy with float glass thereby deriving substantial licence fees from the process. Some commentators, however, have criticized Pilkington for this strategy in the belief that a more offensive marketing strategy might have assured them of a larger market for their own products.

Interstitial strategy

The military strategist seeks his opponent's weakest point to launch his attack. Yet the majority of companies follow a strategy of direct confrontation, launching products into the same markets as their competitors. We have already traced the product life cycle and seen how profit margins are reduced as the innovator faces competition from the followers. How much better it would be for both of them if the late arrivals instead of challenging the innovator on his own territory, had exploited a gap or interstice in his product line.

The computer industry has been dominated by IBM. Other companies have unsuccessfully challenged this leadership, some just survive and others have withdrawn after incurring substantial losses; in the case of RCA an estimated

$500 million was invested in computers before the company decided to leave this industry. Does this mean that no other computer manufacturer can survive? An analysis of IBM's strengths and weaknesses show that it is not uniformly strong in every respect. There are gaps in its product line which it is unable or unwilling to fill. One such, is the scientific market which Control Data Corporation was able to exploit for a number of years without bringing itself into a direct confrontation with IBM. Similar gaps in the product range of the major American aircraft manufacturers have been successfully exploited by small European companies, whereas the larger firms which have attempted to mount a direct challenge without the strength of a large home market have failed.

The 'interstitial strategy' thus follows from a deliberate attempt to avoid a direct confrontation by analysing competitors' weaknesses and exploiting them when matched by a strength of one's own.

Market creation

Most new products or processes substitute within an existing market. Occasionally, however, the chance arises to create an entirely new market where an advancing technology offers an opportunity to develop a product which is entirely new. Thus the TV game was not a substitution and the first company to exploit this innovation was free from direct competition. Where this can be achieved high profits margins may be generated without any great risk. Unfortunately these opportunities are comparatively rare.

'Maverick' strategy

Sometimes the attributes of a new technology severely reduce the market for the product to which it is applied. The market leader is thus vulnerable to the new technology and is unlikely to introduce it since he will be badly hurt as a consequence. This provides an opportunity for the outsider who has nothing to lose but possesses an expertise in the technology.

The French manufacturer BIC has successfully exploited this strategy in developing its range of cheap throw-away products such as ballpoint pens and razors. The marketing of these products had little appeal to the established manufacturers who rightly regarded the alternatives as a threat; nevertheless, they were eventually forced to follow the leadership of BIC and develop similar products.

A 'maverick' strategy enables the application of a new technology in which a company has expertise to launch new products in someone else's market where the innovation reduces the size of the total market. This diversification through technology gives an initial advantage but only succeeds in the long term if supported by an offensive strategy which enables the retention of technological leadership.

Acquisitions – people

An alternative to buying-in a competitor's technology through a licence agreement is to acquire his key staff or even a complete project team. Many companies may regard an aggressive attempt to recruit personnel from a competitor who wishes to retain them, as either distasteful or damaging to their reputation, although they may be prepared occasionally to entice them, perhaps employing a 'head-hunting' agency as the intermediary. This is an ethical consideration each company must decide for itself.

Occasionally, however, recruitment opportunities occur for staff the competitor does not wish to retain either because he is discontinuing a project or reducing his investment in R & D. This can provide an excellent chance to acquire expertise and experience at minimum cost. For example, one British electronic manufacturer was able to recruit the whole of a team who had been working for several years on the development of a new television receiver when their employer decided to discontinue the project. Such opportunities occur only infrequently and no firm could base its strategy upon such chance events. Nevertheless, knowledge of one's competitors can enable a rapid response to change in his policies leading to the acquisition of his unwanted human technological assets when they fit one's own needs.

Acquisitions – companies

There are a number of reasons why one company should decide to takeover or merge with another. Synergy and the matching of resources often play an important part in the industrial logic for a takeover. Technology can be a critical factor.

We have seen that small firms are often highly creative, entrepreneurial and offensive in their strategies. But their mortality rate is also high. The escalation of R & D costs after most of the development problems have been solved may prove beyond their financial resources. Or a new product is successfully launched but the company lacks the production or marketing skills to exploit it fully. On the other hand, the large company may be reluctant to devote scarce resources to an offensive strategy. Thus it makes sense for it to marry its financial resources to the technological assets of a smaller firm which has taken the initial risks.

These are strong reasons why the large company should consider seriously its acquisition policy, not as the occasional exploitation of an opportunity but as a considered act of strategic policy. Several large American companies have been reported as adopting this strategy. Borden, for example, analysed the technological potential of 200 small firms when it decided to expand its polystyrene foam business.

Joint ventures

It is becoming increasingly difficult for any company to obtain all the resources required to develop competitive products, particularly in high technology industries. There will be deficiencies in finance, technological knowledge or market expertise. Thus in recent years there has been a rapid growth in cooperative agreements and joint ventures both nationally and globally. This trend is likely to accelerate in the future.

In many industries the cost of developing a new product is prohibitive even for the industry leaders. Partners must be found to spread the development cost. This is particularly true of the aerospace industry where for many years it has been beyond the ability of any one company to develop a large new aircraft solely from its own resources. The solution may be to form a consortium, as with the European Airbus, or by the subcontracting of major parts of the design and subsequent production to another company, a policy favoured by the US aerospace industry. In aero-engines the three major manufacturers, Rolls-Royce, GEC and Pratt & Whitney have a complex series of joint developments between themselves and with smaller companies whilst remaining highly competitive with each other on other projects.

Nor can any company hope to keep abreast of all the advances in a rapidly advancing technology. The knowledge must be shared. Petrella [11] shows the complex web of global partnerships in the automotive, telecommunications and electronics industries. Figure 2.12 illustrates the situation in the electronics industry in 1989; within two years this had changed considerably due to additional linkages and acquisitions. Another alternative is for the partners to found new joint venture companies as with Living Technology formed between Pilkington and BOC in medical lasers to exploit their own expertise in different technologies and market knowledge.

Frequently a firm will have an expertise in a technology which would enable it to develop products for a market in which it has no experience. It may then be desirable to form a link with another company which can supply what it lacks. Hence ICI found it desirable to form a joint company (Marlow Foods) to develop biotechnology-based food products with RHM which had the strength in the food market. Their first product Quorn, a mycoprotein, was launched as a savoury pie filling in 1987.

Analysis of technological opportunites compared with a capability audit will reveal the existence of a gap between the potential and a company's ability to exploit it alone. It requires creativity and technological knowledge to identify and assess where a strategic opportunity might exist. An evaluation of the technological strength of possible partners is also required. Thus although the creation of a joint venture is a corporate decision it must be founded on the knowledge that only R & D is likely to possess in detail. Adopting a proactive approach based upon a systematic analysis enables the forging of arrangements with the most suitable partners before others take pre-emptive action. Although the best known examples are from the major high technology industries, this is a strategic option that no company can afford to neglect.

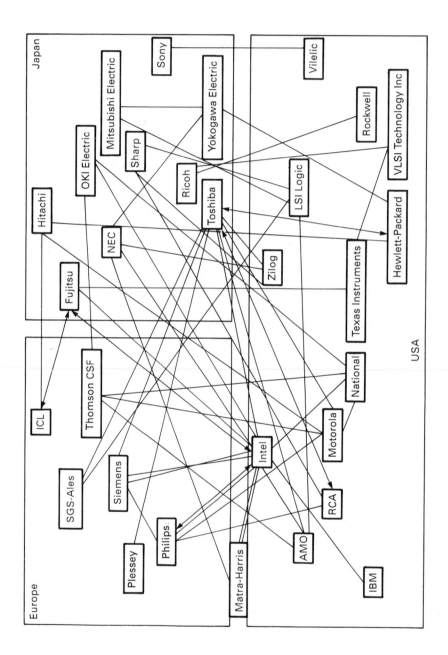

2.12 Strategic partnerships in the electronics industry: Europe, Japan, USA
Source: Petrella, R., 'The Globalization of Technological Innovation' [11]

The industrial life-cycle

The characteristics of the industry life-cycle were described in Chapter 1. It was noted that in their early years change is rapid and product life-cycles are short. As the industry becomes established the pace of innovation slackens as growth is sustained by an increase in the size of the market. With the approach of maturity the market growth rate also slackens, competition increases, and profitability falls. This explains the low returns on investment earned by mature industries such as steel.

The R & D strategy needs to be revised to match the changes required by these stages of industrial growth. In the initial phase success comes from innovation and the pursuit of an offensive strategy. But as product life-cycles extend some resources should be transferred to maintaining or increasing market share. Finally, with the onset of maturity the emphasis shifts to cost reduction programmes and process R & D (Table 2.4). It must be stressed, however, that

Table 2.4 *Relationship of R & D strategy to stage of industry development*

Stage of industry development	Emphasis of R & D strategy			
	Offensive	Defensive		Licensing
	New products New technology	Product improvement	Process improvement	
Phase 1 Rapid growth Low competition	High	Low	Low	Low
Phase 2 Market growth Increasing competition	Medium	High	Medium	High
Phase 3 Maturity Low growth High competition	Low	Medium	High	Medium to high

we are referring to the strategic emphasis. The float glass example shows that occasionally there are opportunities for an offensive strategy even in a mature industry where for several decades most advances have related to minor cost reducing process improvements.

This evolution through the life of an industry imposes severe strains on the organization for R & D. The change in strategic emphasis must be reflected in the selection of projects, the allocation of resources and the recruitment of personnel. But since the industry may take decades to move from one stage to another the implications are frequently overlooked. Thus it is still possible to find companies devoting their effort primarily to an offensive strategy when the industry has reached maturity and the probability of making a successful radical innovation has become very low.

Multi-product companies – portfolio analysis

As with all planning techniques portfolio analysis should not be applied mechanistically. One of its greatest benefits is that it provides a visual representation which assists in the strategic debate. It may, however, direct attention to traditional areas which may have little future potential, a point made by Hamermesch [12]. The wisdom of withholding investment from mature activities must always be questioned when they form a major part of the business. It does, however, assist in the achievement of a balanced portfolio, particularly where technological strength can be associated with a high market potential.

Multi-product companies are likely to have products positioned at different stages in their life-cycles. Additionally the company's competitive strength is likely to vary between products. It is convenient to analyse the overall portfolio by plotting on a matrix. The classical representation is the familiar Boston grid where business growth rate is plotted on one axis and relative competitive position on the other. In recent years there have been a number of variations, a 3×3 matrix now being used by many companies (Hofer and Schendel [13]); these can be adapted to indicate the appropriate R & D strategy to be pursued (Fig. 2.13).

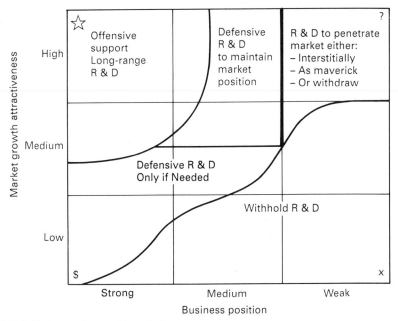

2.13 *R & D strategy in a multi-product company*
Source: Frohman, A. L. and Bitondo, D. 'Coordinating Business Strategy and Technical Planning' [14]

Implementation of an R & D strategy

Although the concept of offensive and defensive strategies is not new, there are few companies which formulate an R & D strategy with the same care as they do their corporate strategy. We must, therefore, ask whether it is either possible or realistic to do so in practice.

The validity of the main elements of the process – environmental forecasting, analysis of the competitive position, technological audit – cannot be doubted. Although each demands the exercise of subjective judgment, they are essential if R & D management is to have an informed view of all the factors bearing on their decisions. The systematic collection of information and its orderly presentation cannot fail to improve the quality of decision-making. Thus it seems reasonable to expect R & D management to undertake these analyses in order to discharge their responsibilities in the best interests of their companies.

There is greater argument about the extent to which these analyses can be structured into a formal planning system. In a period of rapid change formal written plans are probably of little value, since new opportunities invalidating them appear at frequent intervals. Yet these opportunities make essential some form of strategic reference points if they are to be properly evaluated. The growing acceptance of formal corporate planning will impose upon R & D directors in the future a greater requirement for the formal planning of their work. Realism and flexibility must, however, be retained. An R & D strategy must be regarded as a framework which helps to structure decision-making and not as a 'strait-jacket' which imposes unrealistic constraints.

A planning system is only a means to an end. The end we have postulated is better R & D decision-making towards the fulfilment of corporate objectives. Little can be achieved without effective communication between R & D management and top management. The system itself will not ensure this, but it can provide useful assistance.

The alignment of an R & D activity to the strategy of the business is much more complex than the simple analytical approaches described in this chapter might suggest. Every stage of the strategy formulation process is beset by business and technological uncertainties, much of the analysis is based upon judgment, and ongoing programmes provide a momentum which limits the freedom to make sudden changes in direction. Rarely, if ever, is it possible to draw up a strategy *ab initio*. There will always remain constraints which make it difficult to introduce in practice the strategy which might appear desirable. Nevertheless, these problems do not invalidate the arguments in favour of making a conscious effort to derive a strategy for R & D. These benefits can be summarized as:

1. Forcing a rational analysis of environmental trends and the organization's competencies.
2. Encouraging a long-term orientation.
3. Enabling the matching of R & D effort to corporate objectives.
4. Revealing where a technological potential might indicate the desirability of modifying the corporate strategy.

In the final analysis the value of the output is the extent to which it indicates where change is needed. Occasionally this might be a major change in the allocation of resources. More often it results in a gradual change over several years. For the strategy formulation process is not a one-off exercise. It must be a continuous activity. The benefits are likely to be derived as much from the learning acquired as from its reflection in formal plans and action programmes.

Some of the early enthusiasm for detailed formal plans has moderated somewhat in recent years as attention has been focused on the human problems of implementation. We may conclude, therefore, that while the long-term trend will inevitably lead to a closer involvement of R & D decision-making with corporate planning, we cannot expect to find the answers in simple mechanistic formulae. We must seek the correct interplay between the planning system, the process and the people.

Strategy and the R & D director

Formal planning has become a feature of corporate life with which the technologist must learn to live. It can be assumed that technology is an important factor to be considered in the formulation of strategy for companies which invest in R & D. Furthermore, there is need for the corporate strategy to be considered explicitly in the selection and evaluation of project proposals. Thus we have a two-way process whereby:

1. The R & D director makes the technological inputs to the strategy formulation.
2. The corporate strategy provides a framework for R & D decision-making by means of an R & D strategy.

This process can be observed in a few of the most successful technology-based companies – one might be tempted to attribute this success to a planning system which fully integrates technology. Unfortunately, there are still many companies where the R & D director is regarded as a specialist; all too often plans consist almost entirely of financial data in which R & D appears solely as a cost centre. The R & D director is, however, uniquely placed to make a major strategic contribution in three ways. First, his knowledge of technology enables him to identify advances not only in the company's existing technology but also those which could become important in the future. Secondly, he is accustomed to thinking in longer time scales than most of his colleagues. Thirdly, he is able to interpret the implications of information technology throughout the organization. To do so, however, he must willingly embrace this wider role and prepare himself to discharge it.

Strategy *v* entrepreneurship

In Chapter 1 we stressed the importance of the individual, the project champion or entrepreneur. In this chapter we have been concerned with rational analysis. Are these compatible? In recent years there has been much criticism of strategic management as exemplified by the writing of Peters and Waterman [15]. This has been a response to many planning systems which have been over-rigid and have constrained the initiative of individuals, nowhere more important than in R & D.

Both viewpoints are valid. The successful company must combine the two approaches. Strategy provides a framework to give a sense of purpose and cohesion to corporate activities; it is rational and essential. But within that framework it is individuals, often highly emotional, who use their creativity and initiative to provide the means of achieving the corporate objectives and occasionally to change them. The R & D director has the responsibility to manage the interplay between these opposing forces.

Summary

The strategic thinking decribed in this chapter is based upon rational analysis and is becoming increasingly common in large companies. However, readers will have noted that the presence of a coherent strategy is only one of the factors identified in Chapter 1 as contributing to successful innovation. Strategy formulation implies a strong emphasis on *top-down* planning whereas research into successful innovation indicates the need for a *bottom-up* orientation. These two requirements can be reconciled if an innovative management consciously attempts to weld together these two streams of communication. Moore [16] reporting on research into new product development concludes that: 'Only a minority of the new product ideas studied came directly from a formal strategic plan, but most were consistent with it.' This may indicate that there are effective informal linkages even when the more formal approaches we have described are not used. However, there are strong grounds for the view that many companies do not achieve a sufficiently common purpose between their corporate and technological strategies. This is supported by the research of Szakonyi [17] who reports that fewer than 10% of a sample of 170 US companies he studied had effective coordination between R & D and business planning.

The Japanese approach is very different. They invest a substantial effort in determining long-term (25 to 30 year) technological objectives [18]. Few financial constraints are applied to the technologists responsible for achieving these objectives. Their companies are technology and growth rather than financially driven. There is no doubt that this is one of the factors contributing to their industrial success. Readers would, no doubt, wish to work within a similar environment. In its absence the methods advocated in this book should be applied effectively whilst recognizing that this is not the only way of viewing the dynamics of a technology-based company.

Formal corporate planning has now become an important feature of top management policy-making in a wide range of business organizations. The stimulus for this has come from the significant impact that environmental change has in creating the conditions for continued success of most businesses. Those most affected are frequently in technology-based industries, and much of the change is itself brought about by technology.

In strategy formulation an analysis is made of the trends taking place in the social, economic, political and technological environments. This analysis leads to an identification of the business threats or opportunities they are likely to give rise to. There are usually a number of alternative strategies which might be feasible but the optimum will only emerge by comparing what might be possible with the organization's ability either to meet the threats or exploit the opportunities. The strategy instills a sense of purpose in senior management decision-making and should be a major consideration in deciding the level of investment in R & D. It also facilitates the formulation of coherent and consistent guidelines for operational management. But failure to translate the corporate plan into decision-making at all levels will nullify its purpose. Yet there is a great deal of evidence to suggest that this occurs frequently, particularly in relation to investment in technological innovation.

Much of the work undertaken in R & D does not lend itself to detailed planning. This fact must be recognized. The planning system must permit full exploitation of serendipitous discoveries, when appropriate, if it is to be realistic. It was suggested that the difficulty this raised might be overcome if it were recognized that many companies are really engaged in two businesses − one defined by the conscious process of strategic planning, the other based upon the commercial exploitation of random discoveries.

The role of an R & D strategy as a means of relating critical technical decisions, particularly in project selection, to the corporate plan was discussed. The formulation of an R & D strategy demands the same care and analytical approach as a corporate strategy. This involves analysing environmental trends of which technology forecasts are an important element, and a systematic assessment of the capabilities or 'technological capital' of the R & D department.

The characteristics of alternative strategies − e.g. offensive, defensive, licensing, interstitial, 'maverick' acquisitional and joint ventures − were examined. It was concluded that it would be appropriate for most companies to pursue a combination of these strategies although resource limitations would necessitate an emphasis being placed on one or two of them. The strategy manifests itself in the resource allocation decisions made within the R & D department and provides a framework for project decision-making.

Too much emphasis should not, however, be placed upon the formal procedures of planning. The main contribution is likely to be made by the dialogue between R & D and corporate management arising as a result of the planning process.

References

1. Ansoff, H. Igor. *Corporate Strategy*, McGraw-Hill, 1965. Reprinted Penguin Books, 1982.

2. Porter, M. E. *Competitive Strategy*, The Free Press, 1980.
3. Whitaker, R. Reported in *The Financial Times*, 10 Feb. 1970.
4. Kay, N. M. *The Innovating Firm*, Macmillan, 1979.
5. Baker, M. J. *Industrial Innovation: Technology, Policy, Diffusion*, Macmillan, 1979.
6. Hertz, D. and Thomas, H. *Risk Analysis and its Applications*, Wiley, 1983.
7. Ward, E. P. *The Dynamics of Planning*, Pergamon Press, 1970.
8. Durand, T. 'R & D "Programme-Competitiveness" Matrix: Analysing R & D Expertise Within the Firm', *R & D Management*, **19**, No. 2, April 1988.
9. Jewkes, J., Sawers, D. and Stillerman, R. *The Sources of Invention*, 2nd edn, Macmillan, 1969.
10. Cooper, A. C. 'Small Companies Can Pioneer New Products', *Harvard Business Review*, Sept./Oct. 1966.
11. Petrella, R. 'The Globalization of Technological Innovation', *Technology Analysis and Strategic Management*, **1**, No. 4, 1989.
12. Hammermesch, R. 'Making Planning Strategy', *Harvard Business Review*, July–Aug. 1986.
13. Hofer, C. W. and Schendel, D. *Strategy Formulation: Analytical Concepts*, West Publishing Co., 1978.
14. Frohman, A. L. and Bitondo, D. 'Co-ordinating Business Strategy and Technical Planning', *Long Range Planning*, **14**, No. 6, Dec. 1981.
15. Peters, T. and Waterman, R. H. *In Search of Excellence*, Harper & Row, 1982.
16. Moore, W. L. 'New Product Development Practices of Individual Marketers', *Journal of Product Innovation Management*, **4**, No. 1, March 1987.
17. Szakonyi, R. 'Coordinating R & D and Business Planning', *Technology Analysis and Strategic Management*, **2**, No. 4, 1990.
18. Goodridge, M. and Twiss, B. C. *Management Development and Technological Innovation in Japan*, Manpower Services Commission, Report No. MC28, 1986.

Additional references

Bemelmans, T. 'Strategic Planning for Research and Development', *Long Range Planning*, **12**, No. 2, Apr. 1979.

Buggie, F. D. *New Product Development Strategies*, AMACOM, 1981.

Coombs, R. and Richards, A. 'Technologies, Products & Firms' Strategies', *Technology Analysis and Strategic Management*, **3**, No. 1, 1991.

Ford, D. 'Develop Your Technology Strategy', *Long Range Planning*, **21**, No. 3, 1988.

Littler, D. and Wilson, D. 'The Evolution and Strategic Management of

New Technology-Based Sectors: The Case of Computerized Business Systems', *Technology Analysis and Strategic Management*, **2**, No. 2, 1990.

Myers, P. W. 'Non-Linear Learning in Large Technological Firms: Period Four Implies Chaos', *Research Policy*, **19**, No. 2, April 1990.

Souder, W. E. 'Promoting an Effective R & D Marketing Interface', *Research Management*, **XXIII**, No. 4, July 1980.

Teece, D. J. *The Competitive Challenge: Strategies for Industrial Innovation and Renewal*, Ballinger, 1987.

Twiss, B. C. and Goodridge, M. *Managing Technology for Competitive Advantage*, Pitman, 1989.

3

Technology capture, creativity and problem-solving

Change is not a sideline in the business of leadership, it is integral to the whole idea: to describe a man who left things exactly as he found them as a 'great leader' would be a contradiction in terms. A leader may change the map of Europe, or the breakfast habits of a nation, or the capital structure of an engineering corporation: but changing things is central to leadership, and changing them before anyone else is creativeness.

Antony Jay

Technological innovation was described in Chapter 1 as a conversion process whereby scientific or technological knowledge is linked to the satisfaction of customer needs. A concept can originate at either end — by the recognition of a technological potential and developing products based upon it (technology push), or from the identification of a market need that might lend itself to a technological solution (market pull or problem-solving). In both these processes there are two important elements — acquisition of the technology itself and an act of creativity. Whilst the first of these may appear a natural function for any technologist and the second a spontaneous act of creativity, these are an inadequate foundation for a vital corporate activity. The processes must be understood and managed insofar as this is possible, whilst recognizing that they both rely heavily upon individual initiative and insight.

Technology capture

We have already discussed the importance of establishing a sound knowledge base. But how is this to be done? At one time in the past a company was likely to be concerned with only one technology in which new knowledge was generated in a few centres of international repute. A competent technologist could be expected to keep in touch with all the significant developments taking place. Today it is much more complex. Advances in a number of technologies, often in combination, can contain the seeds for opportunities to be exploited or potential threats to be faced. For each technology there are numerous research centres, universities and industrial organizations globally where the key advances might be made. Their findings appear in an ever increasing number of journals but much of this information for a fast moving technology may already be out of date by the time it is published. The problem is too complex and its solution too important to be left to chance. Some mechanism is needed to ensure that the possibility of missing an important advance is minimized without the process becoming too costly.

There are four stages in technology capture:

1. Identification of sources of knowledge.
2. Using these sources.
3. Evaluation of the significance of the knowledge.
4. Acquisition and transfer into the company.

Identification of sources

The sources of knowledge arise outside the company and reside in experts and publications, of which the former are likely to be the most important. However, because the number of potential sources is so great it becomes virtually impossible to establish and maintain contact with them all. Some way of reducing their number whilst still ensuring a depth of coverage is required.

In every field there are a few internationally recognized experts who are aware of, or have contacts with, all or most of the people or establishments where significant work is being undertaken. First, these people must be identified and then a way must be found for utilizing their expertise.

All new knowledge eventually permeates down the information chain. But the aim must be to acquire it at the earliest opportunity so that it can be used to gain a competitive advantage — what we termed earlier as 'just-in-time technology'. In order to do this one must gain access to its source either directly or through an intermediary. Thus the secret of success lies as much with whom you know as what you know. This can only be accomplished by a policy of encouraging activities outside the firm — attendance at conferences, meetings of learned societies, links wth universities and serving on policy committees. An innovative organization should encourage these contacts since it is the most effective way of identifying and building a relationship with the key international experts.

Publications and data banks are also a valuable source of knowledge and have the advantage that they can be accessed from within the company. They are, however, a poor substitute for personal contact.

Using the sources

The value of the key experts is that they are the nodes for information networks in their own fields of expertise and are well placed to recognize the most significant developments as soon as they occur. The aim of the company must be to identify these experts and then to form its own network incorporating them thereby gaining indirect access to these networks. This can be done in a variety of ways.

A personal relationship with the experts on an informal basis is the easiest and the least expensive method of achieving this. But to be really effective the company representative must be of a calibre and have a technical reputation such that there can be a meaningful interchange of information, within, of course, the constraints of commercial security. Enlightened companies recognize that they must give in order to receive. Close contacts with academics in leading universities can also be valuable not only in gaining access to their own work but in using their knowledge of what is being done elsewhere. Some firms attach so much value to this that they have sited their own laboratories close to such centres of learning. A manifestation of this is the growth of science parks to ease the transfer not only between university and company but also between companies working in the same field. This leads to concentrations like that of electronic companies in Silicon Valley in California.

An extension of this process is the establishment of what Perrino and Tipping [1] call a 'global network of technology' which they describe thus:

> This model consists of a network of technology core groups in each major market – USA, Japan, and Europe – managed in a coordinated way for maximum impact. Only a handful of companies including IBM, Ciba-Geigy, Bayer and ICI are pursuing this approach. Since building a technology network can take from 10 to 20 years or more, the companies that move in this direction today will have a clear winning edge tomorrow.

More formally, it may be desirable to pay a consultancy retainer fee to some of the experts who may also be invited to conduct the occasional in-house seminar on recent advances.

By these means it is possible to build up a network of contacts who can perform the role of knowledge gatherers and preliminary filters. They should be drawn from fields which are not only of immediate concern to the company but also those which may have significance in the future.

Technological gatekeepers
The procedure outlined above must be established on a formal basis. Recognition must also be given to the value of random information gleaned by anyone who has outside contacts. There is, however, a tendency to overstress the importance of these contacts for all R & D workers. Allen's [2] research at MIT into communications in R & D indicates that the path by which outside information is transferred into an organization is often indirect and involves relatively few people. He writes:

> These key people, or 'technological gatekeepers', differ from their colleagues in their orientation towards outside information sources. They read far more, particularly the 'harder' literature. Their readership of professional engineering and scientific journals is significantly greater than that of the

average technologist. They also maintain broader ranging and longer term relationships with technologists outside their organizations. The technological gatekeeper mediates between his organizational colleagues and the world outside, and he effectively couples the organization to scientific and technological activity in the world at large.

Most laboratories need to maintain an awareness of advances in several technologies and Myers [3] has shown the presence of specialist gatekeepers for each category of scientific or technological information.

The 'technological gatekeeper' or 'information star' is only one aspect of good communications. There is a wealth of information in every laboratory much of which remains untapped. Fischer and Rosen [4] in a survey of a large US federal government laboratory revealed the existence of people they called 'latent information stars'. One of the conclusions of their study was that where these were basic scientists they saw no rewards for sharing information, an activity which they believed detracted from their personal productivity. Information which is not communicated is largely wasted. All members of a team must be encouraged to contribute information to the common pool. Although the organization and the reward system are important it is the R & D manager's responsibility to play an active part in ensuring the good flow of information.

Evaluating the knowledge

Knowledge in itself is of little value until its significance is appreciated. This is not a problem when it relates directly to the core technology. It can also be a valuable source of data for technology forecasting.

We have seen, however, that a feature of innovation is the association of developments in previously unrelated technologies. Possession of the knowledge, whilst an essential first step, does not guarantee that the linkage will be made. This demands creativity, informed judgment, insight and an ability to understand the potential of the relationships. It can only be done by a person with interests in and an appreciation of the whole spectrum of technologies although it does not require a depth of knowledge in each of them. Such a person might be termed a 'general technologist' analogous to the general manager who can take a holistic view. The R & D director might be expected to be a general technologist but there may well be others who can take this broad perspective.

Technology monitoring

The acquisition and evaluation of knowledge is a continuous process. A possibility for an innovation might be revealed but a piece of the jigsaw may be missing; this prevents the immediate initiation of a development project. One of the constituent technologies might not yet have achieved a sufficiently high performance or a scientific breakthrough may be necessary. Once this has been achieved a practical development is feasible; until then it is not. Thus there is need to store those ideas of potential value, record their possible significance and ensure that there is a good early warning system to pick up new knowledge which would enable them to be realized. A formal system is required to overcome the vagaries of the human memory.

Bright has proposed monitoring the environment on a systematic basis. He writes [5]:

> Note that monitoring includes much more than simply 'scanning'. It includes search, consideration of alternative possibilities, and their effects, and a conclusion based on evaluation of progress and its implications. The feasibility of monitoring rests on the fact that it takes a long time for a technology to emerge from the minds of men into economic reality, with its resulting social impacts. There are always some identifiable points, events, relationships, and other types of 'signals' along the way that can be used in an analytical framework. If a manager can detect these signals, he should be able to follow the progress of the innovation relative to time, cost, performance, obstacles, possible impacts and other considerations. Then he will have two more important inputs to his decisions:
>
> *(a)* Awareness of new technology and its progress;
> and
> *(b)* Some thoughtful speculation about its possible impact.

The monitoring process is based upon a journal in which significant events are recorded. Bright suggests four column headings: 'Date', 'Event and Technical Economic Data', 'Possible Significance', and 'Things to Consider'. The events may be likened to the pieces of a jigsaw. Over a period of time, the number of pieces we have increases. Not all may be useful, but by careful selection and assembly, a picture of one part of the future may slowly emerge. Sometimes a vital piece may be missing; this could lead to the deduction − 'if advance X were made, then innovation Y would be possible' with the corollary 'we must monitor the environment carefully for signs of X being achieved since we shall then know that Y is feasible'. The reporting of X might appear first in an obscure scientific paper in any part of the world.

The attraction of monitoring is that it can be performed by any individual manager for his own information. It is surprising how much one person can glean from systematically processing the information received daily. The richness of the information and the deductions are obviously enhanced if organized on a departmental or interdepartmental basis.

Acquisition and transfer into the company

Most of the new knowledge acquired will be within the areas of expertise of the R & D staff. The main problem arises when it relates to an area where that expertise does not currently exist. It must then be acquired from outside.

Where the need is temporary it is possible to subcontract the work to a university, contract research organization or to employ a consultant. This may be the only economical way to solve the problem but it is not ideal. The outside agency lacks the commitment to the company's objectives and may have no sense of urgency; it may be difficult to specify the needs precisely; communication problems will occur; and there may be a low sensitivity to the organizational and practical problems in converting that knowledge into a product within the company.

If the need is likely to be long term and the cost is not excessive it is usually much more desirable to build up the expertise in-house. The most effective way is to recruit high calibre individuals or teams on a permanent basis. There is a price to pay and difficulties can arise in fitting them into a rigid payment system since the persons required may be unwilling to accept the salaries the company can offer within its payment structure. However, these problems must be overcome if this essential human resource is to be acquired. The needs of innovation are of greater importance than conformance with a personnel policy. Flexibility is required. The absence of this flexibility in many large mature companies is another reason why they are so often less innovative than smaller competitors.

Discussion

Knowledge is the foundation for all technological innovation. It is too important to be left to chance and the initiative of individuals although their contribution cannot be ignored. Too many companies lack an adequate system for analysing and capturing new technology effectively. It must be stressed that the competitive advantage is obtained by the first company to apply effectively new knowledge in its products. Timing is of the essence of success. Figure 3.1 shows how the factors identified in this section can be brought together within a management system for technology capture.

Creativity – a widespread need

Every successful innovation starts from a creative idea, whether it be a new application for a technology, the satisfaction of a new identified need or of an old need in a new way, or the manufacture of a product by a different method. Some organizations appear much more adept at generating these ideas, while others devote substantial resources to producing results which are at best mundane. But there is no simple recipe leading to the attainment of a creative organization, for the processes involved are imperfectly understood. Nevertheless, it is possible to identify some of the ingredients – creative people, an environment receptive to new ideas, and the use of creative problem-solving techniques – the conditions for which are largely within the R & D director's control.

Creativity does not lend itself to planning. Ideas often seem to appear spontaneously as if by a flash of inspiration from within the deepest recesses of the brain when the mind is consciously engaged elsewhere. The classical story

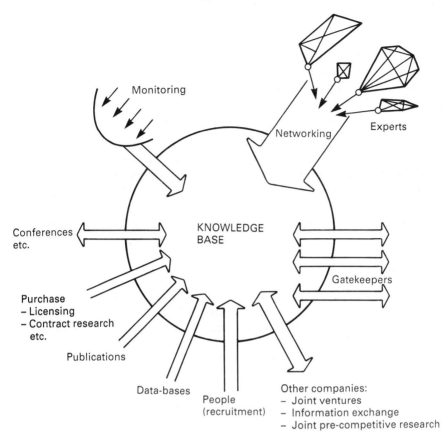

3.1 Managing the knowledge base

of Archimedes in the bath is but one example. Then there is the serendipitous discovery arising from the observation of an unplanned event and noting its significance. But it was not just random good fortune that made Newton and Fleming household names. Is it possible to understand and reproduce the mental process that leads from the observation of a phenomenon to a theoretical hypothesis or the idea for its practical application? Certainly not in our present state of knowledge and probably never. Would other people have drawn the same inferences as Fleming if they had observed the chance killing of a bacterial culture on a specimen dish, or would Fleming himself have reacted in the same way on another occasion? We do not know, but it seems unlikely that any managerial act can increase the chances of such occurrences.

And yet the example of penicillin is remarkable in that Fleming did not realize its practical application as an antibiotic − this development was left to others a decade later. Why was this obvious step overlooked for so long? Under the right stimulus or in the right environment penicillin antibiotics could certainly have appeared earlier.

This inspirational view of creativity is challenged by Weisberg [6] who maintains that: 'Scientists do not make great intuitive leaps into the unknown independently of what has come before, but that even in its most impressive manifestations, scientific discovery develops incrementally and is firmly based on the past.' He illustrates this view of creativity with a detailed analysis of the events which led to the discovery of the structure of the DNA molecule by Watson and Crick and Darwin's theory of natural selection. He concludes:

First, the early theorizing, based relatively directly on ideas of others, was shown to be inadequate in several ways. These inadequacies were addressed by modifying various parts of the old theory, and the result was a new theory. These modifications did not occur all at once in a leap of insight, but rather involved a series of small accommodations as each difficulty was handled. There seemed to be no one point at which everything fell into place, but rather, there was a gradual closing in on the final theory as bits and pieces fit together. On the whole the process is remarkably similar to the process of creative problem-solving.

Although Weisberg is here concerned with scientific discovery rather than technological innovation there is a strong parallel. Three critical elements he notes are the acquisition of knowledge, painstaking experimentation and patience.

However, it should not be thought that creativity is only required for the original concept of an innovation. There is ample scope for the application of creativity, although of a lower order, at every stage of development. Most industrial projects are concerned with solving specific problems. There is usually a number of alternative solutions to each of these problems. Some of the alternatives will be barely adequate, others expensive, some highly promising, but rarely one which is undoubtedly superior to all the others. It is essential that the first alternative thought of should not be proceeded with as is so frequently the case, without stopping to generate and evaluate a wide range of possibilities, for it is rare for first thoughts to provide the most creative solution.

Thus the need for creativity in R & D appears to extend to all working levels. To judge from the views expressed by research directors, many of the graduates they employ lack the necessary imagination and insight in spite of their ability to perform highly programmed complex tasks well. Thomason [7], in contrast, maintains that much tension and frustration amongst creative scientists in industry is caused by their employment on tasks where there is little opportunity for the exercise of their creative ability. There are, of course, various degrees of creativity as indeed there are tasks which vary from the routine to those where imagination can be exercised to the full. Only a few highly creative persons can be provided with sufficient opportunities for exercising their creative ability within an industrial organization. But this is only one aspect of the problem which extends right through the R & D department utilizing people of varying creative capacities in a variety of tasks demanding different degrees of originality.

Thus while recognizing that chance plays a major part and the highly creative act cannot be planned, we may conclude that action can and ought to be taken by the R & D manager to ensure:

1. The amount of creative ability in his organization at all levels is adequate.
2. The creative potential of his staff is identified.
3. The opportunity for the exercise of creativity in each job is analysed.
4. Tasks and people are matched so far as possible.
5. The creation of a working environment in which:
   (a) Unplanned creative ideas are received with an open mind and are not rejected out of hand because they do not accord with current plans or conventional practice.
   (b) Creative solutions within ongoing projects are encouraged, particularly in the early stages when exhaustive searches should be made to ensure that the subsequent investment of time and effort is well placed.

The creative process

Most people use the word 'creativity' quite freely without pausing to think about the process by which the act or idea they describe as creative came about. Our judgment is largely subjective, coloured by the extent to which the solution has surprised *us* because it was not what *we* would have expected. Nevertheless, it is possible to identify a number of features frequently found in those ideas we would describe as creative.

Many definitions of creativity stress the part played by the imagination in generating new concepts or unusual solutions to problems. Schon [8], for example, describes invention as 'a non-rational process'. Other writers widen the definition to embrace analytical techniques based upon systematic search methods such as morphological analysis. Whether the latter can properly be called a creative process is of minor importance to practical technologists concerned more with the results than the means by which they are achieved. In his analysis of twentieth-century inventions and innovations Richardson [9] identified four sources of innovative knowledge and the creative spirit – inspiration, investigation, experimentation and the search for an economic niche – more than one of which may be present in a particular example.

Whether the idea be generated by imagination or by analysis, *patterning* appears to be an important element. Facts, relationships or parameters, both technical and non-technical, are arranged in new combinations. Examination of important innovations frequently shows that they resulted from the bringing together of a number of individual advances, often in unrelated fields. The apparently steady path of progress typical of many technologies, which we note when discussing technological forecasting, supports the view that most new ideas do not result in the overthrow of the existing state, they merely extend it. Thus the majority of creative acts do not represent great leaps forward and in retrospect may appear to have been the only logical steps to have taken. Yet when two or more independent advances are associated for the first time, either intuitively or as a result of systematic search, the result of this technological synergy can appear as a great surprise to those working in the area. Kuhn has put forward the theory that the progress of science is generally evolutionary punctuated by occasional revolutions. During the evolutionary periods there is a tendency to ignore facts which do not fit the established theory. Thus the belief that the world was flat was sustained for many centuries, evidence to the contrary being rejected until Copernicus was able to gain acceptance for a new theory, namely that the world was round.

New patterns often emerge from the *transfer of knowledge* from one field of activity to another. The value of multi-disciplinary teams in associating concepts developed in one technology to the problems arising in another has been stressed in the literature. Unilever uses multi-disciplinary teams extensively. The use of phosphates or citrates to inhibit crystal growth on teeth leading to the development of anti-tartar toothpastes was suggested by a technologist from the detergent business. Individuals with a wide range of technological interests can also provide the catalyst which bridges the barriers between technologies. Bradbury [10] quotes how an aerodynamic solution, the application of spiral strakes to distillation columns, provided an elegant solution to stability problems where conventional stress analysis had focused on reducing their resonance by increasing their inertia. This illustrates the importance of the general technologist role. Frequently, however, in an industrial setting, the greatest advances have been made across industry boundaries. Thus 'innovation by invasion' results from the application of a technological development in one industry (e.g. chemicals) to the problems of another (e.g. textiles). The identification of such opportunities often results from a creative act.

Many ideas result in a *redefinition of the problem* and vice versa. Sometimes it is the insight enabling this redefinition which represents the creative thought, but often it is only after the new solution has been found that earlier abortive efforts are seen to have been concentrated on an imperfect understanding of the problem. Once the problem is redefined the solution often appears so obvious that it is difficult to understand why it was not thought of long before. Sir James Taylor [11] quotes an interesting historical example:

(Creativity) is now, and always has been, inhibited by conventional wisdom and established habits. For example, the horse-drawn chariot, which included that fantastic invention, the wheel, was almost certainly used 4,000 or 5,000 years BC which was before anyone had the simple idea of riding the horse. A complicated solution of a problem often precedes a simple one.

The surprise element of so many creative ideas arises because they *do not result from the exercise of logical thought processes* and conventional wisdom. When our minds are concentrating on a problem they are focused and constrained within self-imposed boundaries. Thus much of the accumulated knowledge and experience available to us is excluded, and may only be brought to bear on the problem when the restraints are relaxed, when we allow 'our minds to wander'. The modern highly trained technologist, however, is a specialist educated to apply scientific logic. The development and exercise of his imagination form no part of his training and may even be actively discouraged. He is expert at applying his knowledge to effect modest improvements upon the status quo, but it is extremely difficult for him to project his mind forward and to think normatively. Moreover, Hudson's [12] research suggests that the personality of the average scientist, who is a convergent thinker, is basically non-creative. One of the reasons for the effectiveness of the multi-disciplinary team is that it brings together people working within different mental constraints. An extreme case of this comes from a large American research organization where one of the most creative members is a former theologian. Inevitably many of his ideas cannot be translated into practical terms, but, occasionally, however, he does come up with a proposal which would not have resulted from the normal thought processes of his technological colleagues and yet proves to be technically feasible.

Closely akin to the 'non-rational' and the 'subconscious' is *fantasy*. In this state the mind moves into new worlds of the imagination untramelled by the limitations of the physical world as we know it. Technologists may be inclined to reject the idea of fantasy as a factor they should take into account when solving practical problems. And yet it does seem to be an important element in much creative thinking. Bellman [13], for example, in a history of mathematics states that in the development of new concepts, '. . . the pressures of reality breed fantasy which astonishingly turns out to be a new reality'. Gordon [14], goes further and draws upon the wish-fulfilment theory of Freud* to develop a 'fantasy analogy' as part of the synectics problem-solving technique.

Although we may never know exactly how new ideas are generated, studies of creative acts lead us towards some understanding of the processes at work. In particular we have noted the following:

- New combinations or patterns of existing knowledge and concepts springing from the imagination or resulting from techniques of systematic analysis.
- Association of ideas, often from widely different spheres of learning, which enable new patterns to emerge.
- Creative solutions arising from or resulting in a redefinition of the problem.
- Mutual stimulation between persons of different intellectual backgrounds.
- A freeing of the mind from the constraints of normal logical rational thought processes.
- The role of fantasy in achieving a state of detachment.

Any understanding of the creative process is, however, little use to us as managers unless it is possible to build practical problem-solving techniques upon it. Later in this chapter it will be shown that the first tentative steps have been taken. In particular, the techniques of synectics and lateral thinking owe a great deal to this understanding and have proved useful in a wide variety of applications.

* 'Success depends upon his ability to defer consummation of the wish in fantasy and to make real the wish by embodying it in a work of art', by Gordon, W. J. J., *Synectics* (p. 48).

Creative individuals – main characteristics

Any attempt to match a person's creative ability to the needs of the tasks he is required to perform is of little avail in the absence of means for evaluating the individual. Various methods which attempt to assess and measure creative potential have largely failed to provide a convincing or practical guide for the manager. In the present state of knowledge we must conclude that decisions must be made without precise and meaningful measures. Thus we must fall back upon subjective judgment based primarily upon an assessment of the individual's past creative achievement. This is a function of two factors, hereditary and environmental, which cannot be separated clearly in any evaluation of performance. In spite of the difficulty, research studies do give us some help in determining some of the characteristics we might expect to find in creative people, though it must be recognized that these can only be broad outlines at best.

Creativity is closely associated with the ability to ask the right questions. But the creative person may or may not be capable of providing the technical solutions. Thus a high level of formal education is not necessarily a prerequisite. Edison and Faraday, for example, had no formal education. This is equally true today of many successful entrepreneurs in technology-based companies where the founder may not be highly qualified in the technology (e.g. Alan Sugar of Amstrad). Many good ideas can originate outside the R & D department although it is only the technologist who can translate them into products. In one company the idea for a new baby-care product came from the chairman's wife. The R & D Director believed the proposal to be impracticable but nevertheless felt bound to initiate a project; this culminated in a practical solution which eventually became a highly successful product. Technological arrogance and the 'not-invented-here' syndrome can be major barriers to the exploitation of good ideas originating from outside R & D.

Not surprisingly the research supports the view that creative people are intelligent, although some studies indicate there is little correlation between creativity and intelligence above an IQ level of about 120. Research at the Institute of Personality Assessment and Research (IPAR) into the creativity of several groups of people, particularly architects, leads to the same conclusion. MacKinnon [15], reviewing the IPAR studies, states that they 'suggest that we may have overestimated in our education system the role of intelligence in creative achievement'.

MacKinnon describes the picture of the creative individual emerging from these studies in the following terms:

> The evidence is clear: The more creative a person is the more he reveals an openness to his own feelings and emotions, a sensitive intellect and understanding self-awareness, and wide-ranging interests including many which in the American culture are thought of as feminine.

and:

(The results) suggest that creative persons are relatively uninterested in small details, or in facts for their own sake, and more concerned with their meanings and implications, possessed of considerable cognitive flexibility, verbally skilful, interested in communicating with others and accurate in so doing, intellectually curious and relatively disinterested in policing either their own impulses and images or those of others.

Gerstenfeld [16], summarizing the work of a number of research studies, lists the main indicators of creative performance as: the intelligence threshold, previous creative acts, a commitment to internal standards, and persistence.

Is it possible to measure these factors and use the results in personnel selection? The psychologists Cattell and Drevdahl [17] have identified sixteen personality factors* which they score based on ten to thirteen items for each factor (16 PF test) [18]. In one series of studies using the 16 PF test the scores for leading research scientists (chosen by their professional societies) were compared with science teachers and scientific administrators. The results showed (with a quantitative measure) that the researchers were more schizothyme, self-sufficient, emotionally unstable, bohemianly unconcerned and radical, than the teachers and administrators and also, though less uniformly, more dominant and paranoid and lower on compulsive super-ego (will control). These tests provide a profile for the typical researcher which can be used as a basis for assisting in the selection of personnel. Although the work of Cattell and Drevdahl dates from the early 1950s there is little evidence the 16 PF tests have been used to any extent in R & D personnel selection.

The manager concerned with the level of creativity in his own organization would like to identify the creative achievement of his present staff. He would also wish for criteria to help him in personnel selection. In assessing his current staff he has a certain amount of factual evidence such as publications and patents to add to his subjective judgment. Combined with the judgment of others, and against the background of the findings of MacKinnon *et al.* he should be able to obtain a fair evaluation of their past creative performance. But it is doubtful whether many managers do attempt to make a systematic critical evaluation of this important characteristic of their staff. Rickards and Besant [19] recommend a creativity audit.

The manager's task in personnel selection is more difficult. Most, if not all, the information usually available to him relates to the candidates' past academic achievement. But since the educational system makes little or no attempt to evaluate creative potential the prospective employer is likely to have little to guide him.

---

* Cyclothymia (i.e. tendency to alternating excitement or depresson) *v* schizothymia (i.e. dissociation of the emotional from the intellectual life); intelligence; stability; dominance; surgency (enthusiastic, heedless and happy-go-lucky); super-ego; adventurous cyclothymia; sensitive emotionality; paranoia; Bohemianism; free-floating anxiety; radicalism; self-sufficiency; super-ego will control; psychosomatic anxiety.

Creativity in innovation — the manager's problem

Creativity in the organization results from a delicate interplay of many factors. This is seen by referring to Fig. 3.2 which shows how some of the most important influences interact to form a closed system.

It is not in itself sufficient that an R & D department contains persons of creative potential employed in tasks where they can exercise their creativity. The potential has to be released. This will occur only if staff are motivated to submit creative ideas with the knowledge that the organizational climate is receptive to them. In the absence of this knowledge they will, after a time, either cease to make creative proposals or alternatively leave the organization for another where they consider their ideas will have a greater opportunity of acceptance. The receptivity of the organization to new ideas is also likely to stimulate their generation in other ways. It may for example lead the organization to experiment with creative problem-solving techniques. Although not all ideas will be accepted or, if accepted result in successful innovations, those innovations that do occur will further stimulate the desire to propose imaginative solutions. Indirectly a successful reputation for innovation will attract creative people to the organization.

Thus we have a positive feedback system with success breeding success. Conversely, the system may well work in reverse when lack of encouragement drives out creative people, ultimately resulting in an organization where innovation becomes virtually impossible irrespective of the resources devoted to R & D. At the centre of this system lies the R & D director. He determines the working environment where creativity is encouraged. Yet he has to maintain a delicate balance between the unorthodoxy of this creativity and the stable progress of established projects requiring, as they do, the maintenance of schedules and attention to detail. Ultimately, of course, it is top management who determine the climate for creativity and innovation throughout the company. Without their support the research director can do little, and it is they who are responsible for his own appointment. It is generally true that top management gets the research director it deserves, thus reluctance to innovate or accept new ideas at the top is likely to be reflected in the character of work of the laboratory at all levels.

Organizational environment

We have seen that the research director's role extends beyond the mere recruitment of creative people and hoping for the best. Yet it is difficult, if not impossible, to prescribe how he should manage his laboratory to achieve the delicate organizational balances necessary for a creative environment. Nevertheless, it is not so difficult to identify features of the organization likely to inhibit the development and dissemination of creative ideas. Some of these barriers can be removed.

A great deal can be learnt from studying successful organizations. From these

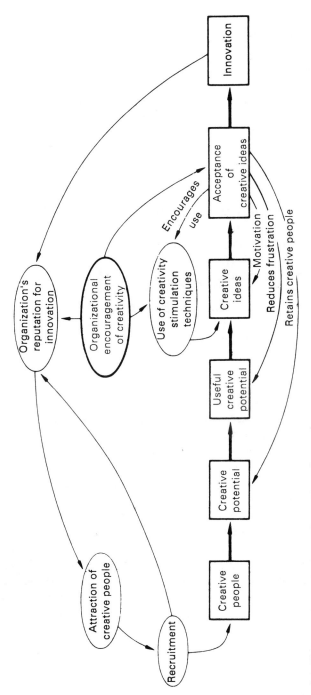

3.2 The creative organization – interaction of many factors

one can gain an understanding of the characteristics they commonly exhibit. This assists in forming a clearer appreciation of what we should be aiming for in our own situations, though it must be admitted that this may be of little practical help apart from the removal of identified barriers to creativity.

A useful starting point is to examine what creative people themselves think. Some evidence comes from a survey by the Industrial Research Institute, Task Force on Stimulation of Creativity and Productivity [20]. American scientists in industry, universities and government services, deemed by their colleagues to be highly creative, were asked to rank ten factors according to their importance in enhancing the creativity of an organization. The factors, together with the number of times each was listed as one of the two most important items, are as follows:

Freedom to work on areas of greatest interest	44
Recognition and appreciation	41
Broad contacts with stimulating colleagues	38
Encouragement to take risks	26
Tolerance of non-conformity	12
Monetary rewards	8
Opportunity to work alone rather than in a team	3
Creativity training programmes	2
Criticism by supervisors or associates	1
Regular performance appraisals	0

Some care is necessary in interpreting the results of this survey. First, the subjects were all men of high individual creative accomplishment. It might be argued that such people can be given greater freedom to work on problems that interest them personally than can be accepted for the organization as a whole. Secondly, their average age (53) is high and their attitudes are not necessarily representative of those of younger men. The rating given to creativity training programmes, for example, might well have been higher amongst younger men or in those companies, still small in number, with experience of these recently developed techniques. Nevertheless, the great importance attached to the first four items cannot be ignored.

Kaplan [21] lists five characteristics of the creative organization: receptivity to new ideas; pressure to produce; toleration of the 'oddball'; freedom to choose problems and change the direction of research; and incentives for creativity.

The inclusion of 'pressure to produce' may seem to conflict with the need for a high degree of freedom. Kaplan maintains that, although the motivation must come primarily from within the individual, management plays an important role in creating the conditions for these self-imposed pressures. This suggests that creativity is likely to flourish less in a *laissez-faire* environment marked by the abdication of management, than in a participative managerial situation. We shall return to the topic of managerial style in R & D in Chapter 7.

The 'oddball' is defined by Kaplan as a person who does not comply with organizational norms. He is likely to be a loner who does not work well in a team. This is not surprising since the creative person can be expected to produce ideas which challenge the accepted views of his colleagues. It must not be assumed, however, that all creative people necessarily withdraw from close contact with others. The organizational danger is that the isolation of the more creative persons cuts them off from effective communcation with the rest of the team thereby leading to the rejection of many of their ideas. This is much less likely to occur where a ready acceptance of the challenge from new ideas is normal. For the research director the problem is to avoid the withdrawal of his most creative staff from the rest of the department to such an extent that their ideas have no impact. While tolerance of unorthodox behaviour may be the solution this is unlikely to be present in the absence of a working environment which is itself creative.

Gerstenfeld's research (see [16] earlier) isolates ten characteristics of the creative organization: the individual challenge of a specific task, realistic goal setting by R & D management, immediate feedback, a reward system and recognition for creativity, openness and allowance of conflict, cross-fertilization of ideas resulting from a mixture of specializations, job enlargement by following the idea from conception to practical realization, involvement rather than satisfaction, porous organizational boundaries, and less conformity.

These studies bring into relief the delicate task of the industrial R & D manager faced with reconciling the conflicting requirements, on the one hand, of engendering a climate conducive to the generation of creative ideas and, on the other hand, maintaining firm control of the work of his department in the furtherance of company objectives. He must aim to strike a balance between:

1. The provision of freedom for creative individuals to follow their own areas of interest
   and
   The maintenance of interpersonal relationships and the team spirit necessary for the accomplishment of specific projects.
2. The provision of opportunities for multidisciplinary exposure
   where
   Work efficiency often requires teams of people with similar backgrounds and training.
3. The toleration of non-conformity in an industrial setting
   where
   The culture stresses conformity.
4. The mismatch between personal objectives
   and
   Organizational objectives.

These dilemmas can be resolved. The importance attached by Kaplan to the pressure to produce, and Gerstenfeld to the individual challenge of a specific task, suggests that the channelling of creative talent to meet departmental objectives may not be as difficult as is sometimes thought.

Techniques for creative problem-solving

Commercial organizations will always remain partly dependent upon spontaneous creative acts. These cannot be planned although we have seen that it is within the power of management to increase the probability of their occurrence. Their number may be small but occasionally they are of such significance that a company or an industry will be transformed. Important as these rare events are, most industrial activity is concerned with the generation of ideas to meet specific needs or objectives. R & D resources are allocated to find answers to questions such as: How can we satisfy this particular market need? How can this new technology be turned to a profitable application in a product or process? How can we resolve this specific problem?

The answers to these questions normally leave little to chance, but the success of the projects initiated to meet these needs rests largely upon the quality of the ideas or concepts upon which they are based. Management nowadays can draw upon a variety of operational techniques whose purpose is to improve problem-solving capability, particularly in the generation of creative solutions [22].

The remainder of this chapter will be devoted to brief descriptions of the techniques available. They fall into two distinct categories:

1. *Analytical* – these apply logical thought processes exercised within a formal structuring of information.
2. *Non-analytical* – these stimulate imaginative thinking along unorthodox paths, deliberately aiming to free the mind from the constraints imposed by logical analytical thought processes.

Analytical problem-solving techniques

Analytical techniques are designed to reveal new approaches to problems or new combinations through systematic search. This is possible because of the ordered nature of the world we live in. Once we can establish the fundamental relationships linking what previously appeared as random facts, we can apply those same relationships to extend our knowledge still further to make new discoveries. This is the essence of the scientific method. For example, by relating the properties of the elements to their atomic weights Mendeleyev created the Periodic Table and used a simple morphological matrix in the process. This provided a framework of analysis which he then used to predict the existence and properties of elements, unknown at the time, such as scandium, gallium, neon, krypton, etc.

Similar methods have been developed to solve management problems. The techniques of management science and systems thinking depend upon our ability to establish fundamental relationships which can then be applied to solve practical problems. Many of these techniques such as decision trees, heuristics and model building can be applied to solving the type of problem met within

R & D, though often their full potential cannot be realized without recourse to that less easily defined ingredient – imagination. Sometimes the techniques merely serve to provide the framework which channels the imagination.

No attempt will be made to give the reader a comprehensive list of the standard management science techniques which on occasions could be helpful to the R & D manager. We shall, however, consider a few techniques which are particularly appropriate in the solution of R & D problems.

Attribute analysis

We shall examine the attributes of a technological phenomenon in Chapter 4 to show how they could be analysed in relation to their suitability in respect of a variety of practical applications. Some of these applications are not immediately apparent, for example, the Flymo lawnmower or a hospital bed for burns patients are not obvious uses for the hovercraft principle. Listing the technological attributes does, however, concentrate the mind on the essentials of the phenomenon, thereby increasing the probability that the imagination will conceive an idea for a practical use.

A similar approach has been developed for analysing the *attributes of a product*. An existing product is analysed in detail relating each part to the purpose it serves. The individual part can then be examined to determine whether it can be improved or used in a different combination to create a new product. Osborn [23] suggests the following questions which should be considered during the analysis – Put to other uses? Adapt? Modify? Reduce? Substitute? Rearrange? Reverse? or Combine?

Morphological analysis

Morphological analysis involves the identification of the main parameters or functions of a problem, together with the various ways of achieving each of them. Every combination represents a possible solution. We can see how this works in the generalized example below (Fig. 3.3). In this problem there are three functions A, B, C; A can be satisfied in three ways, B in two ways, and C in four ways.

Function	*Alternatives*			
A	A_1	A_2	A_3	
B	B_1	B_2		
C	C_1	C_2	C_3	C_4

Total number of possible combinations $\Sigma \, ABC = 3 \times 2 \times 4 = 24$

3.3 Morphological matrix

Analysis shows that in this case there are 24 different combinations which might satisfy the problem. In a practical example some of these combinations will already be known, some could be rejected as non-feasible, but some might reveal new possibilities not previously thought of. Figure 3.4 gives part of a morphological matrix for room heating. The ringed path would represent a coal-fired central heating system using pumped hot-water distribution and radiators. Examinations of this particular example show that it covers systems as widely different as the open fire to underfloor electric heating. It is left to the reader to see whether the matrix can be used to find a feasible potential new system from the $5 \times 2 \times 6 \times 4 = 240$ possible combinations.

Bridgewater [24] shows how morphological analysis was used in the OSCAT project for examining alternative process routes from an iron-bearing feedstock to a shaped pure iron product. In this example there were 387, 420, 489 possible routes for six stages and a computer was used in the preliminary selection process. The large number of options identified can often present a major practical difficulty. Industrial companies that have based new products on the use of this technique are AKZO (surface coatings), Ciba-Geigy (dyes) and 3M (laminates).

Morphological analysis is often described as a technique for technology forecasting because of its ability to suggest future technological advances. However, most of the practical applications have been in the identification of possible design configurations and as such it may more properly be regarded as a technique for generating new ideas for the solution of practical technological problems.

Needs research

There are a number of ways in which the needs of the user can be analysed systematically to yield ideas for new products. Attribute listing is but one of them.

Needs research is a term applied to the analysis of complex technological systems incorporating a large number of subsystems. Rarely if ever is the total system replaced. Advances are made by the improvement of subsystem performance. It is vital in a problem of this nature that R & D investment is devoted to that part of the total system where it will bring the ultimate user the maximum benefit. Needs research is a modelling technique which enables a thorough investigation of the total system performance, in relation to forecasted user needs, to be made in respect of alternative investments in improvements of the subsystems which comprise it. This approach has been applied to the British telephone network. This is essentially a long-term problem where the investment in the existing telecommunications infrastructure is so great that it cannot be replaced as a whole. Even a subsystem improvement may have a gestation period of 5–10 years with a further period for installation and a long working life ahead of it. The needs research team, a multidisciplinary group, uses technological forecasting techniques to forecast the user needs the total system should aim to satisfy including non-technical factors such as cost to the user and level of usage which have a bearing on the suitability of technical alternatives.

A theoretical model of the system is then constructed. This enables a detailed

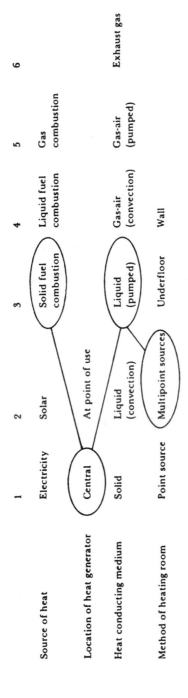

	1	2	3	4	5	6
Source of heat	Electricity	Solar	Solid fuel combustion	Liquid fuel combustion	Gas combustion	
Location of heat generator	Central	At point of use				
Heat conducting medium	Solid	Liquid (convection)	Liquid (pumped)	Gas-air (convection)	Gas-air (pumped)	Exhaust gas
Method of heating room	Point source	Multipoint sources	Underfloor	Wall		

3.4 Partial morphological matrix for room heating

analysis of the effect upon the total system, under a variety of conditions of new components or improved subsystems. Thus it is possible to identify where changes can yield the greatest benefit, provided they can be proved technically feasible at an acceptable cost. This could not be achieved without the model because the complex interrelationships within the system can reduce the effectiveness of improvements to subsystems in relation to the whole.

The important features to note in needs research are:

1. The analysis starts with the *user's needs.*
2. The use of *technology forecasting* to forecast these needs at some future period.
3. The employment of a *modelling technique* to relate subsystem and system performance to those needs.
4. The experimental determination of where investment can yield the best results *before a project is defined.*

Non-analytical aids to creative problem-solving

Analytical methods gain ready acceptance both by technologists due to their scientific training and other managers increasingly accustomed to the use of quantitative techniques and formal planning. But the needs of the non-rational creative process are imperfectly understood and lead to a great amount of misunderstanding.

In the years following the war, investment in R & D became sufficiently large to cause attention to be focused on formal evaluation and control techniques. Little sympathy was shown for the view that R & D was a creative process which could not be controlled rigidly. This argument was often regarded as special pleading by those who were attempting to avoid the financial disciplines considered essential elsewhere in the business. Some of these suspicions were well founded. Not infrequently the demands of creativity were used as an excuse for neglecting essential aspects of good managerial housekeeping.

In recent years a realization has grown that formal control methods do not alone hold the key to R & D effectiveness. More attention is now being paid to the very real problems of striking a correct balance between creativity and control. This change in attitude has been brought about by two developments. First, several companies which had moved farthest in the application of formal managerial control systems became concerned by what appeared to be a reduction in the quality and quantity of the output from their R & D departments, in spite of rapidly increasing expenditure. Secondly, the work of Gordon and Osborn in the USA and De Bono in the UK suggested that the stimulation of creativity could be advanced beyond the stage of managerial abdication. Several major companies began to experiment with these problem-solving techniques on both sides of the Atlantic. We have now reached a stage where these techniques are gaining increasing acceptance, although their use is still not yet widespread.

A parallel can be drawn between creativity training and management education. Until the 1940s it was widely assumed that managerial ability was inborn and little could be done to develop it. Admittedly, a few American business schools, notably Harvard, had been in existence for a quarter of a century but their influence was limited, particularly in Europe. The teaching methods used at that time were primarily situational based on case studies of real-life company problems. In the past twenty years management education has moved into a second phase following a rapid expansion. More clearly defined concepts are emerging, thereby adding a theoretical base to complement the situational pedagogy which still retains an important role.

With creativity, it is only more recently that people have begun to accept that, although hereditary differences are important, it is possible to develop methods for generating creative ideas. As yet the methodology may appear cumbersome but it has been shown to yield practical results by a significant number of major companies, for example, Unilever and ICI in the UK and Kimberly-Clark and Singer Sewing Machine in the USA. These successes give strong support to the view that these techniques represent more than a passing fashion. As their use spreads they may be expected to attract further research leading to refinement of the techniques themselves and a deeper understanding of the underlying principles. Thus creativity can be left less to chance and can be brought into the orbit of organizational thinking.

An interesting feature of these developments is that the majority of the workers in this field possess a background in technology rather than in the behavioural sciences, although their thinking has been influenced by the earlier work of psychologists. These technological origins account for the majority of applications being made in R & D, although they are now spreading into other areas of business, such as the generation of ideas for strategic change, diversification or acquisition.

The creative problem-solving techniques aim to develop the conditions noted earlier as characteristic of the creative process, within an experiential situation. Consequently, both the most widely used methods, synectics and lateral thinking, share much common ground. Before examining how the techniques are applied we shall now take a brief look at the principles underlying them.

Some underlying principles of non-analytical methods

Are we solving the right problem?
Normally we do not question the validity of the problem we are asked to solve. It may appear self-evident; it may have been defined by a superior, perhaps as a project definition; or it may emerge as a problem area arising during an R & D programme. But frequently the solution we would regard as creative is novel and unexpected because it derives from a redefinition of the problem, sometimes suggested by the solution itself.

An example will illustrate what so often happens in practice. A problem arose in the manufacture of glass due to erosion of the channels carrying molten glass from the furnaces caused by the impingement of gas bubbles upon the refractory lining. A research project was initiated which eventually solved the problem. In this solution the molten glass was kept out of contact with the upper surface of the lining by creating a chamber of pressurized gas which absorbed the released bubbles. The sidewalls of this chamber remained in contact with the molten glass, and were coated with platinum. This solution was considered satisfactory and glass furnaces were modified. However, some time later it was suggested that the difficulty could have been resolved simply by sloping the channels to allow the gas to escape to atmosphere without impinging on the lining. The research project had solved the problem 'How do we protect the refractory lining?' whereas the later solution answered the question, 'How can we remove the gas bubbles?'

A fixation on the perceived problem makes it extremely difficult to see it differently. With the benefit of hindsight it is easy to be critical and suggest that a trained mind should be capable of looking at a problem from all angles, without recourse to a technique for creative problem-solving. But we know from hard experience that the 'obvious' solution resulting from a redefinition often comes after we have expended a great deal of effort on 'solving' the wrong problem.

Can we break our logical thought patterns?

All our thinking is constrained by our past experience and training. Faced with a new problem we try to apply old solutions. This can be seen when a business problem is given to managers from different areas of activity, say an accountant, a production manager or a marketing manager. Working alone each will apply his own filter to the information he is given, selecting those items which lie within his own field of competence. These will then be ordered and processed according to the specialist tools of his trade. In this way we all apply a subjective distortion to our understanding of a problem which we proceed to solve by extrapolation along a narrow path.

Many of the techniques of technology forecasting are based upon a normative approach where the mind is cast forward to some postulated state in the future. It is then necessary to trace a feasible path back to the present. A similar approach is required in creativity in order to escape from the constraints of our traditional thought patterns. But this cannot be achieved easily.

The mind seems incapable of making a quantum jump consciously, however hard we try. It must be enticed away from logical thought. The main contribution of the creativity spurring elements of the synectics technique is the development of a contrived procedure whereby the mind is slowly drawn away from the perceived problem to a state of detachment. This 'metaphorical distance' is later reduced with the objective of thereby finding a new point of entry to the problem. In this process, fantasy, metaphor or analogy are deliberately used within a structured procedure aimed to open up the whole scope of a person's stored experience, to make it available for solving the problem, rather than the restricted field to which logical thought invariably limits one.

The use of fantasy, metaphor, and analogy in creative thought is not new. Many examples occur in the history of science. Analogies from nature are particularly frequent. John Smeaton, for example, is supposed to have obtained his design for anchoring a lighthouse to the Eddystone rock by studying the root formation of the oak tree. What is new, however, is the development of techniques and procedures for creating deliberately what previously happened by chance.

Can we become more receptive to new ideas?
Everyone tends to reject a new idea particularly if it is put forward by someone else. When presented with a new idea we seize upon its weaknesses and tend to disregard its advantages. Edward de Bono demonstrates this human characteristic by showing a radical new design of a wheelbarrow to a group of people. His proposal has both merits and weaknesses, but the reactions of the group are almost invariably negative, concentrating on the failings of the design to the almost total exclusion of its advantages. The objections to a new idea are often trivial, based on some minor point of technical feasibility. No idea is likely to be fully developed when it is first propounded. It needs an opportunity to advance and mature. But this development can be prematurely frustrated unless judgment is deferred. Thus, one of the aims in creative problem-solving is the attainment of a mental climate which is prepared to accept ideas with an open mind so that the creative process can continue without hindrance. Practical difficulties must be shelved temporarily and evaluated at a later stage.

In this context the reader with no experience of creative problem-solving techniques should examine his own reactions to them. Experience with groups exposed to them for the first time produces varying responses. Where they have been presented to groups who have attended the sessions in order to explore the potential of the techniques they have been readily accepted. In group exercises they have produced novel solutions to problems which individual members of the group have previously worked on without success. In contrast, the identical presentations have been rejected when incorporated as part of a general management training programme. The response has been sceptical, creative solutions were not forthcoming and the initial prejudices of the group were confirmed.

The author's own experience reflects the difficulties on first encountering these techniques. Initial scepticism on reading Gordon's (see [14] earlier) work was only partially relieved by the knowledge that *synectics* had been successfully applied by several large companies. Conviction of its real value only came from the opportunity to observe a consultant working on a practical assignment with an industrial company. Since that time experience of a number of groups has led to an increasing realization of the considerable potential these techniques possess as aids in solving practical problems. Because these techniques are largely experiential and emotional rather than cognitive, a true appreciation can only be gained from participation with an open mind.

How can others help?

The value of interdisciplinary groups has been referred to earlier. Unlike the Delphi process we are not seeking an objective analysis of the problem for, as Taylor (see [11] earlier) says, 'Creativity is an emotional activity and is inhibited by working within fixed rules.' This activity can be stimulated by the free ranging interplay of minds from different backgrounds working together within the same experiential setting. The cross-fertilization and mutual support from group working is of such value that it has been considered essential by many workers in the area. Richardson [9] (see earlier) analysed the major discoveries, inventions and other innovations of the twentieth century and concluded that: 'The preponderant majority of the discoveries/inventions are almost never the result of the work of one individual working alone.' This suggests that an important managerial role is to act as the catalyst and the creator of an environment to stimulate the free interchange of ideas.

There are, however, practical difficulties with group working in an industrial company. There must be a nucleus of people who wish to participate in group problem-solving. There may also be difficulties in gathering them together when required. This has led to increased attention being paid to the development of techniques the individual can use by himself. Nevertheless, group interaction and the multi-disciplinary approach are of such value that individual techniques are never likely to replace group working entirely.

We shall now examine how the principles discussed above have been incorporated within the framework of a practical technique. The three techniques used most widely are:

1. Brainstorming.
2. Synectics.
3. Lateral thinking.

Brainstorming

Brainstorming is a completely unstructured approach to group problem-solving. A group of people gathered together interact to generate ideas spontaneously. No evaluation is permitted during the brainstorming session, and, although no attempt is made to relate the ideas, they owe a great deal to mutual stimulation and cross-fertilization. The proceedings are recorded for subsequent detailed examination.

The value of brainstorming in solving technical problems is generally thought to be limited. Rickards [25, 26], however, reports that about half a sample of twenty brainstorming sessions he investigated in an R & D environment produced utilizable results. The most common application of the technique has been to generate ideas for new products, notably consumer products, and in the creation of advertising slogans.

Brainstorming can be associated with both synectics or lateral thinking. De Bono [27] explains the procedure of a brainstorming session in his book *Lateral Thinking for Management*.

Synectics

Synectics is a word coined by Gordon (from the Greek: to join together different and apparently irrelevant elements) to describe the structured group technique he devised for achieving a creative problem-solving climate. It aims to achieve: freedom from constraints imposed by the problem as stated, elimination of negative responses, deferred judgment, and escape from the boundaries imposed by orthodox thought patterns.

Gordon started his research into the creative process in 1944, while at Harvard University, by observing groups working on problem-solving tasks. By recording and subsequently analysing what went on in these groups, he gained an understanding of the processes sufficient to enable the development of an operational technique which he later applied to the solution of practical problems while working in the Invention Design Group of Arthur D. Little Inc. Since 1960, when he formed Synectics Inc, the operational methodology has been further refined.

The operational technique was based-on the assumptions that:

1. The creative process in human beings can be concretely described . . .
2. The cultural phenomenon in invention in the arts and in science are analogous and are characterized by the same fundamental psychic processes.
3. Individual process in the creative enterprise enjoys a direct analogy in group process.

Synectics is based upon the theory that the probability of problem-solving success is increased by an understanding of the emotional and irrational components of the creative process which are considered more important than the intellectual and rational elements.

The problem-solving process, sometimes called an 'excursion', consists of the following main stages:*

1. A statement of the problem as given.
2. Analysis of the problem.
3. Immediate suggestions or 'purge'.
4. Statement of the problem as understood.
5. Increasing the 'metaphorical distance' by means of:
   (a) direct analogy,
   (b) personal analogy,
   (c) compressed conflict.
6. Possible repetition of 5 in a different context.
7. Fantasy force fit.
8. Generation of possible solutions, or 'viewpoint'.

The *problem as given* is a statement of the problem as it appears at the outset.

* For a fuller description of the rationale underlying the technique illustrated with transcripts from actual problem-solving sessions, see Prince, George, *The Practice of Creativity* [28].

This initial understanding of the problem as we have seen already, may represent inadequately the fuller perception of it from which the solution will emerge.

A multi-disciplinary group usually contains several members who have no previous knowledge of the problem. During the *analysis of the problem* the person who has brought the problem to the group gives them a detailed description and answers any questions arising from his presentation. At this stage of the process some *immediate suggestions* for a solution will occur. It is important that the excursion should pause while all these ideas are tabled. The name *purging* is sometimes given to this activity which is designed to free the mind of any preconceived ideas likely to frustrate the subsequent process unless cleared out of the way before proceeding further. The purge ideas are recorded for later evaluation since they may be worthy of detailed consideration. In some ways this is an unfortunate term for in a carefully selected group first ideas may be important.

The problem can now be defined within a wider context − *the problem as understood*. Several problems as understood may emerge, representing the different perceptions of the problem by individual members of the group. They are recorded but no attempt is made to arrive at consensus. The scene is now set for the *excursion* where analogy and metaphor are used to 'make the familiar strange'.

Direct analogy is used to put the participant into a state of speculation by drawing his mind further from the immediate problem. Analogies drawn from nature, particularly biology, are frequently effective in the solution of mechanical or physical problems. It is important, however, that the mind should explore the analogy in its own right avoiding constant reference back to the problem to be solved, since the purpose of the exercise is to gain distance from the problem.

In *personal analogy* the participant is invited to imagine himself the subject of the analogy and to describe his feelings, e.g. 'Imagine you are a homing pigeon − what does it feel like?'

During the next stage of the excursion the group is asked to generate simple paradoxical phrases drawn from the analogy experience to enable generalization from the specific, e.g. dependable unreliability, enforced liberty, living death. These phrases are called the *compressed conflict*.

The excursion has now reached the period when the 'metaphorical distance' is at a maximum. It is now necessary to travel back to the practical problem slowly, attempting to relate the compressed conflict to the problem as understood, first at the conceptual level, called the *force fit*, and subsequently at the practical level of *suggested solutions* to the problem. The suggested solutions, of which there may be a number, end the creative phrase. Normal analytical methods of evaluation must then follow to reduce the ideas to practice.

Table 3.1 gives the main stages in a practical excursion. Since the whole process took longer than one hour it may not be immediately apparent to the reader how one stage led to another. The eventual suggestions appear obvious in retrospect, but the participating group were convinced that the solutions

Table 3.1 *Example of a practical excursion*

Stage		
1.	Problem as given	Participant stated he would like to solve the problem of how he, an amateur, could build a regular brick wall.
2.	Analysis	He explained the practical difficulties he experienced.
3.	Problem as understood	How to make a novice a consistent bricklayer.
4.	Direct analogy	Homing pigeon. (Reason for choice: a young pigeon has to learn to return to the same place consistently.)
5.	Personal analogy	A participant described his feelings as a homing pigeon.
6.	Compressed conflict	The personal analogy led to the choice of the compressed conflict. Unreliable habit.
7.	Second direct analogy	Sleep was chosen as an unreliable habit. Consideration of dreams led to the second compressed conflict.
8.	Second compressed conflict	Active sleep.
9.	Fantasy force fit	Engage mind on another problem during bricklaying.
10.	Practical force fit	Put patterns on the brick, so that attention is directed to creating a regular pattern.
11.	Suggested solutions	Inscribed datum lines on bricks. Patterns or wavy lines on bricks. Colour coding.

emerging from the idea of getting the bricklayer to concentrate on another task within his capability gave an entirely new insight into the problem.

The inadequacies of a brief written description of a technique, both emotional and experiential, are obvious. Gordon (see [14] earlier) and Prince (see [28] earlier) explain the rationale at greater length, illustrating their description with transcripts from actual excursions. But the best advice to someone new to the technique is to try it for himself in a session conducted by an experienced synectics leader.

During the past few years there has been an increasing tendency for synectics practitioners to stress the behavioural aspects of group interactions. We may expect them to draw closer to the work of sociologists as the techniques develop.

Lateral thinking

Edward de Bono uses the term lateral thinking to describe the characteristics of creative imaginative thought which distinguish it from the normal logical thought process he calls vertical thinking. He has developed a range of techniques to promote lateral thinking in practical problem-solving which have been widely used with success in in-company consultancy and in educational courses.

Both lateral thinking and synectics recognize the same underlying principles of creativity. De Bono [27, 29, 30], however, stresses the importance of patterns in shaping our ideas − patterns which are dependent upon the sequence in which information is received. In order to think creatively it is necessary to break the established pattern to permit the reordering or repatterning of the information to give new insight into the problem, thereby leading to imaginative solutions.

De Bono suggests that before searching for new ideas it is useful to examine current ideas and identify the major influences giving shape to them. The things to look for include: *dominant ideas* which unduly influence and constrain the examination of a problem; *tethering factors* which are included without re-evaluation in every solution to a problem; *polarizing tendencies* which make it difficult to assume a position between two extremes: acceptance of *boundaries* which unnecessarily restrict the area of investigation through preconceptions of where the solution of the problem ought to lie; and *assumptions*.

Examination of the influences constraining the formation of current ideas leads to the development of a methodology for avoiding them. Amongst the techniques he suggests are: questioning – asking why?; the rotation of attention between different aspects of a problem; the forced generation of a quota of alternative solutions; changing the concept; dividing and subdividing the parts of a concept or conversely bridging the divisions between the parts in a new way.

The most interesting techniques relate to the introduction of discontinuity into problem-solving in order to break the established pattern. This can often be achieved by the chance introduction of what at first sight may appear irrelevant. Thus exposure to irrelevancy can be of great value as a stimulant. De Bono uses this in support of multi-disciplinary groups where ideas can come from outsiders with no specialist knowledge of the problem area or from the transfer of knowledge between disciplines. But he regards random environmental exposure equally valuable, e.g. browsing through books or wandering around shops.

Discontinuity can, however, be introduced deliberately. He supports the use of analogy but in a much more direct form than results from the synectic excursion. A simple technique for introducing discontinuity is the random word method in which a word is selected, usually from a dictionary, by a random process. This word is then developed in as many ways as possible in an attempt to establish a link with the problem which may lead to a new insight resulting in a creative solution. Although group use of the random word technique is the most effective, it does lend itself to use by an individual working alone.

The logical thought process consists of sequential steps each of which must be logical in itself. It is familiarity with this process which causes us to pause and evaluate each stage in the development of a creative idea. This we have seen already inhibits the creative process for which it is desirable to defer judgment. De Bono uses the term 'intermediate impossible' to describe an idea which, while wrong or impossible in itself, can be used as a useful stage in the development from a real problem to a logical or rational solution. He coins the word Po which is 'designed to introduce the discontinuity function into thinking to help creative and insight changes' [30]. Like the analogy, Po is an operational mechanism to escape from old ideas or generate new ones.

Examples of how the techniques of lateral thinking have been developed into an operational methodology can be found in the literature. But, as with synectics, a full understanding of how they work in practice can be obtained only from experience. It is hoped, however, that this brief description will encourage the reader to explore more deeply what promises to become a highly useful aid to the solving of a wide range of business problems.

An integrated approach

Creativity is not an end in itself. It is needed both to identify new market opportunities and applications for a technological potential. But it is of no value if the gap between market need and technological potential cannot be bridged. Thus it is essential to ensure that creativity is channelled within a systematic approach to innovation whilst ensuring that the creative individual is not frustrated.

Too much attention in the past has been focused upon the merits of particular techniques in problem-solving whereas what is required is an integrated approach to innovation of which creativity is only one aspect. A number of ways of achieving this have been advanced. Two methodologies which have been applied successfully are needs assessment and SCIMITAR. Holt, Geschka and Peterlongo [31] have developed a technique of systematically analysing existing, future, emotional and rational needs to define product characteristics and specifications involving customers, technologists and marketers. Carson and Rickards' [32] SCIMITAR is a modelling approach combining both analytical and non-analytical creativity techniques into a system covering a project from inception to product launch.

Some organizations have cast their net very wide in a search for new product ideas. In seeking possible applications for a new technology Unilever has involved multi-functional teams as well as consumers, using a range of techniques including brainstorming and lateral thinking. This could generate up to 100 ideas of which about three would have sufficient merit to warrant market research. Similarly, a Swiss chemical company used brainstorming (500 ideas from 80 people), morphology (175 ideas from 30 people) and Delphi (90 ideas from 80 people). These and similar efforts have in common:

1. The involvement of a wide range of people with different backgrounds.
2. The use of a variety of techniques.
3. A systematic approach.
4. The large number of ideas generated to produce one concept for a marketable product.

Summary

The concept for a new product or process arises from the linkage of technological knowledge and ideas for satisfying new needs or existing needs in new ways.

It was seen that the acquisition of the knowledge is so vital to the innovative activity that it cannot be left to chance. It must be planned and managed. Ways in which this can be achieved systematically were discussed. This process can sometimes lead directly to the generation of ideas for exploiting its potential. Alternatively it provides a pool of knowledge available for satisfying creative ideas originating elsewhere.

The knowledge, however, is of no value in isolation. Creativity is the key ingredient which leads to an innovative product. A number of characteristics of creativity were discussed. It must be concluded that this human attribute is insufficiently understood and does not lend itself to formal planning. But this does not mean that it should not be a major concern of R & D management.

The effectiveness of research and development depends upon the quality of ideas. Creativity is needed both in the formulation of project concepts and in the solution of problems arising during development. Thus the need for creativity is widespread.

The role of the research director in providing the stimulus for new ideas is vital. This extends beyond ensuring the recruitment and retention of creative people and their employment on tasks where their creative potential can be utilized. He must understand the creative process and the characteristics of a creative environment. This enables the identification and removal of barriers to creativity by managerial action, although it is not possible to prescribe procedures to relieve him of the burden of using his judgment in striking a balance between the often conflicting requirements of a creative environment and the effective control of projects.

Both analytical and non-analytical techniques are available to aid in problem-solving. Analytical techniques, based on normal logical thought processes, of particular use in solving technological problems are attribute analysis, morphological analysis, needs research, and technology monitoring.

The non-analytical techniques are designed to stimulate the imagination to gain new insight not obtainable through logical analysis. The use of these techniques, particularly synectics and lateral thinking, although not yet widespread, has shown great promise.

A major problem to be overcome is scepticism. This makes it difficult to establish the initial impetus for ideas and innovation, particularly where senior management is not convinced of the validity of the techniques. Organizational factors such as market commitment and the project champion are generally regarded as more important than systematic approaches to creativity. Many managers who have participated in brainstorming sessions also appear to experience difficulty in evaluating the ideas generated. Nevertheless, it is likely that the use of these techniques will become widespread in the future and this will be accompanied by further refinement of the techniques.

Although the importance of the corporate environment and team work has been stressed, the role of the individual in improving his creative ability should not be overlooked. Altier [33] suggests four ways of aiding this process: 'Removal of barriers — stop being your own worst enemy; forget everything you know (the relationships); remember everything you know (the pieces); and rearrange everything you know (same pieces, new relationships).' It will be noted that these factors are an integral part of the more formal techniques we have described in this chapter.

References

1. Perrino, A. C. and Tipping, J. W. 'The Global Management of Technology: A Study of 16 Multinationals in USA, Europe and Japan', *Technology Analysis and Strategic Management*, **3**, No. 1, 1991.
2. Allen, Thomas J. *Managing the Flow of Technology*, MIT Press, 1977.
3. Myers, L. A. 'Information Systems in Research and Development; the Technological Gatekeeper Reconsidered', *R & D Management*, **13**, No. 4, Oct. 1983.
4. Fischer, W. A. and Rosen, B. 'The Search for the Latent Information Star', *R & D Management*, **12**, No. 1, April 1982.
5. Bright, J. R. 'Evaluating Signals of Technological Change', *Harvard Business Review*, Jan./Feb. 1970.
6. Weisberg, R. 'The Myth of Scientific Creativity', Chapter 5 in *Windows on Creativity and Invention* (Richardson, J. G. (ed.)), Lomond, 1988.
7. Thomason, G. F. *The Management of Research and Development*, Batsford, 1970.
8. Schon, D. A. *Technology and Change: The New Heraclitus*, Pergamon Press, 1967.
9. Richardson, J. G. 'The Historic Boundlessness of the Creative Spirit', Chapter 24 in *Windows on Creativity and Invention* (Richardson, J. G. (ed.)), Lomond, 1988.
10. Bradbury, J. A. A. *Product Innovation: Idea to Exploitation*, John Wiley, 1989.
11. Taylor, Sir James. *Proceedings of Conference, 'Luck, Serendipity or Planning'*, The Research and Development Society, 1970.
12. Hudson, L. *Contrary Imaginations*, Penguin Books, 1966.
13. Bellman, R. 'Mathematics, Systems and Society', *Report No. 2 from the Committee on Research Economics (FFK)*, Royal Institute of Technology, Stockholm, 1971.
14. Gordon, W. J. J. *Synectics*, Harper & Row, 1961.
15. MacKinnon, D. W. 'The Nature and Nurture of Creative Talent', *American Psychologist*, **XVII**, No. 7, July 1962.
16. Gerstenfeld, A. *Effective Management of Research and Development*, Addison-Wesley, 1970.
17. Cattell, R. B. and Drevdahl, J. E. 'A Comparison of the Personality Profile (16 PF) of Eminent Researchers with that of Eminent Teachers and Administrators, and of the General Population', *British Journal*

of Psychology, **46**, 1955.

18. Cattell, R. B., Eber, H. W. and Tatsuoka, M. M. *Handbook for the Sixteen Personality Factor Questionnaire (16 PF)* (1970 edn), Institute for Personality and Ability Testing Campaign, Illinois, 1970.
19. Rickards, R. and Besant, J. 'The Creativity Audit', *R & D Management*, **10**, No. 2, Feb. 1980.
20. Parmenter, S. M. and Garber, J. D. 'Creative Scientists Rate Creativity Factors', *Research Management*, Nov. 1971.
21. Kaplan, N. 'Some Organisational Factors Affecting Creativity', *IRE Transactions on Engineering Management*, Mar. 1960.
22. Van Grundy, A. B. *Techniques of Structured Problem-Solving*, Van Nostrand Reinhold, 1981.
23. Osborn, A. F. *Applied Imagination*, Charles Scribner, 1963.
24. Bridgewater, A. V. 'Morphological Methods: Principles and Practice', in: *Technological Forecasting* (Arnfield, R. V. (ed.)), Edinburgh University Press, 1969.
25. Rickards, T. *Problem Solving Through Creative Analysis*, Gower, 1974.
26. Rickards, T. 'Brainstorming in an R and D Environment', *R & D Management*, **3**, No. 3, June 1973.
27. De Bono, E. *Lateral Thinking for Management*, McGraw-Hill, 1971.
28. Prince, G. M. *The Practice of Creativity: A Manual for Group Problem-Solving*, Harper & Row, 1970.
29. De Bono, E. *The Mechanism of Mind*, Jonathan Cape, 1969.
30. De Bono, E. *Practical Thinking*, Jonathan Cape, 1971.
31. Holt, K., Geschka, H. and Peterlongo, G. *Need Assessment*, John Wiley, 1984.
32. Carson, J. W. and Rickards, T. *Industrial New Product Development: A Manual for the 1980s*, Gower, 1979.
33. Altier, W. 'A Perspective on Creativity', *Journal of Product Innovation Management*, **5**, No. 2, June 1988.

Additional references

Crawford, M. C. *New Product Management*, 2nd edn, Richard D. Irwin, 1987.

Kirton, M. J. (ed.). *Adaptors and Innovations: Styles of Creativity and Problem-Solving*, Routledge, 1989.

Nystrom, H. *Creativity and Innovation*, John Wiley, 1979.

Parker, R. C. *The Management of Innovation*, John Wiley, 1982.

Rickards, T. *Creativity at Work*, Gower, 1988.

Rickards, T. *Stimulating Innovation: A Systems Approach*, Frances Pinter, 1985.

Steiner, G. A. *The Creative Organization*, University of Chicago, 1965.

Whitfield, P. R. *Creativity in Industry*, Pelican, 1975.

4

Project selection and evaluation

If the productivity of research is to be increased, then project selection, budgeting, and control must be placed on a more logical and scientific foundation, and not left entirely to the hunches, intuition, and guesses of individuals and committees. This does not mean that management will be absolved from all planning responsibility; research is the antithesis of certainty, and all that formal systems can do is to increase the ratio of objective to subjective criteria on which plans and decisions are made.

A. Hart

The purpose

Project selection is one of the two most critical and difficult decision areas in R & D management. Of equal importance is project termination due to the high proportion of projects discontinued before their development is completed. The factors to be considered in both selection and termination are almost identical, the major difference being the quality of information upon which the decisions are based. Consequently, the early stages of an R & D programme, leading to the establishment of technical feasibility, can be thought of as an investment of resources to reduce uncertainty, or, in other words, a refinement of the information which determines the viability of the project.

A project generates its own momentum and there is an implicit assumption amongst those working on it that it will be allowed to proceed to completion unless some major new factor emerges. This is perhaps desirable, for the morale of those most intimately concerned with the project would undoubtedly suffer if they were conscious that it was liable to be cancelled at any time. Nevertheless, it must be kept under constant review by the person or group responsible for deciding its fate. Evaluation should, therefore, be a continuous process with the possibility of termination at any time in the light of additional information. It is not, of course, practicable to update all information at frequent intervals. Periodic major re-evaluations are also required when every aspect of the project can be reviewed. These reviews usually precede milestone events such as a major investment decision or they can be organized on a periodic basis, say every three or six months.

The evaluation procedure should therefore be used not only for selection decisions but also as one of the main management control systems within R & D. As with all control systems, it needs to be established on a formal basis, operating as follows:

1. Identification of factors relevant to the project decision.
2. Evaluation of the project proposal in relation to these factors, using quantitative information where available or subjective quantitative judgments, where appropriate, when actual data is unobtainable. It is essential to record all assumptions and quantitative estimates as a control standard for future reference.
3. Selection or rejection of the project proposal on the basis of the evaluation made in 2.
4. Identification of areas where additional information is required and the investment of resources to obtain these data.
5. Comparison of new information arising from 4, with that used in the initial decision, hence the importance of recording the earlier assumptions and estimates.
6. Assessment of the impact of any variance revealed in 5 upon the continued viability of the project.
7. Decision to terminate the project, or to proceed with it, repeating the stages 4 to 6.

The principal factors which have to be considered when establishing an evaluation procedure relate to:

1. The financial benefits expected to be earned from an investment in the project.
2. The effect of the project upon others within the R & D portfolio.
3. The impact of the project, if successful, upon the business as a whole.

Since *financial benefit* is what all commercial organizations are seeking from technological innovation, this would appear to have overriding importance in any evaluation system. Research on both sides of the Atlantic reveals that most companies do, in practice, base their choice of projects almost entirely upon some form of cost-benefit analysis, sometimes modified by subjective probability estimates for technical and commercial success. Although a system which reduces to one figure (i.e. a cost-benefit ratio) expressing the merit of the project has the advantage of giving one simple input into the decision-making process it is only adequate if: (a) the estimates for both benefit and cost are reasonably accurate, and (b) all relevant factors can be expressed in quantitative terms. Rarely are these two conditions satisfied. Nevertheless, quantitative analysis remains an important feature of any evaluation system even if it is not sufficient by itself. The financial aspects of evaluation will be explored in greater detail in Chapter 5.

Project portfolio balance is primarily an internal matter for the R & D department. Nevertheless, it should be an important consideration in project selection, for the ultimate concern of the business is not the performance of individual projects but the continuing contribution to the company's profitability accruing from its total investment in R & D. A project can only be evaluated in isolation from other projects being worked upon or under consideration if there is no mutual interference. Normally this would only be the case in the unusual situation where unlimited resources are available or where the number of potential projects is very limited.

Attention should also be paid to the selection of applied research projects which can contribute to more than one product line. Ruffles of Rolls-Royce [1] describes how the company has achieved what he refers to as 'economy in technology' by concentrating research on areas such as the gas generator and wide chord hollow turbine blades which are applicable to engines of varying size.

A successful project can have a major impact *upon the business as a whole*. The remoteness from commercialization at the time of initial selection, coupled with a relatively low success rate, may tempt the R & D manager to ignore many factors which appear of minor significance to him at the time, but become of great importance at a later date. He should plan for what one research director has called 'the contingency of success'. In our discussions of R & D strategy we saw that a technological innovation of great benefit to one company may have a disastrous effect upon the profitability of another. Apart from strategic considerations, this can arise in many ways. The new product might, for example, be unsuitable for distribution through the existing logistic system, or it might require launching costs beyond the company's financial resources, or it might necessitate an investment in new manufacturing equipment and skills. Many of these factors are difficult, if not impossible, to evaluate quantitatively, although they have a significant influence upon the ultimate profitability of the innovation. If this is so, then some account of them must be taken at the earliest stage of project selection and evaluation.

A project evaluation system is not an end in itself; it is only an aid to decision-making. Brunsson [2] draws a distinction between rationalism and impressionism. Most of the discussion in this chapter relates to the rational systems as distinct from the impressionistic which Brunsson describes as 'consideration of a few conspicuous or at least easily visible characteristics of the project proposal. On the basis of this a picture of the project is created which is perceived as good or bad.' An amalgam of the two approaches is required and this is where the R & D manager applies his experience, knowledge and insight but within a systematic structure. We shall now consider these factors in greater detail.

Project selection

A three-step approach is advocated:

1. The determination of the rationale for the product.
2. Preliminary assessment of its implications.
3. Detailed evaluation.

The rationale for the product

The success of a project is determined in the market-place. The elegance of a technical solution or the use of a new technology are of no value unless they result in a product people are willing to purchase. Too often new products are developed which fail to satisfy this overriding criterion.

There are only two reasons why a customer buys a product. It either performs better than the current offerings or it can deliver a comparable performance at a lower price. If it can satisfy both these criteria, so much the better. Thus the first step in considering any proposal for a new product concept must be a clarification of what it can contribute to the buyer. Unless it is possible to establish a strong linkage to a market need there is no case for proceeding. This is an essential first step in the analysis although the identification of a potential market does not necessarily indicate that the proposal is viable without a more detailed evaluation. It results in either a rejection or a qualified acceptance.

A clear understanding of why the concept may be attractive to the market has important implications for the formulation of the design specification and the development programme. Many detailed decisions will be affected by whether its consumer appeal is based upon its performance or price.

Preliminary assessment of its implications

The next stage in the evaluation is an examination of the total process which would lead from the successful completion of R & D through to the product's use and eventual disposal by the customer. This involves company activities such as product design, manufacture, and marketing and distribution; there may be system and infrastructural needs without which the product would not be viable; and there may be implications for the customers other than first cost and performance. The analysis might reveal weak links which could raise serious constraints to its commercial success or areas which could have repercussions elsewhere in the business. These cannot be ignored and it may be necessary to modify the project concept to accommodate these considerations. Amongst the questions to be asked are:

1. Does the company possess a competence in all the required technologies?
2. Can the product be produced with equipment currently available to the company?

3. Is the product compatible with the company's marketing and distribution system?
4. Is it compatible with other components in the larger technical system of which it forms a part?
5. Is the necessary infrastructure in place?
6. What might be the impact on the customer?

Technological competence
The more radical a proposal, the more likely it is that it will require the application of knowledge and skills new to the company. Their acquisition involves more than the allocation of financial resources. It takes time and may necessitate preliminary applied research to validate the feasibility of the proposal. A sound technological base must be established and there must be a reasonably high level of confidence before a more detailed proposal is submitted.

Production capability
A product incorporating new technology, for example new materials, might be incompatible with current production equipment. As a consequence a substantial investment in capital equipment together with the recruitment and training of staff may be needed. If the company is unwilling or unable to make this commitment there is no point in proceeding with the proposal unless it lends itself to some modification in order to minimize the impact. In some cases it might be desirable to carry out an investigation of these aspects before initiating any work in R & D.

Marketing and distribution
The new product may appeal to markets with which the company is unfamiliar; for example, it might be designed for an industrial market whereas previous products have been for the consumer market. A new drug might be aimed directly at the consumer where existing products have been prescription drugs marketed to medical professionals. The distribution channels might also be different. Both the financial and organizational aspects of these changes must be assessed.

Operational System Compatibility
Many products are parts of a larger technical operational system. Improvements in one component of the system are only of value if they enhance the performance of the total system. An extension of the in-use life of one component, for example, is of little or no value if it extends beyond that of the system itself. Improving the sound reproduction quality of a music centre is pointless if it is beyond the capability of the loudspeakers to exploit. Physical and technical compatibility with the rest of the system is essential. Whilst it is desirable whenever possible to design a new component so that no change is necessitated elsewhere this is rarely feasible. Thus the initial assessment must be carried out not only in relation to its effect on system performance but also

to take into account the modifications needed in other parts of the system in order to incorporate it and the likelihood of their acceptance.

The customer may purchase a complete system or use it in conjunction with other equipment he possesses. Thus in developing a new system for audio recording (e.g. DAT or mini CD) it is not sufficient to ensure that there is equipment on which it can be played. A view must be taken of whether the recording companies will adopt the new format and whether the user will be prepared to make the additional personal investment, both of which are to a large extent interdependent. The availability of software may be a critical element in the acceptance of a new hardware system.

Infrastructure

A new product may be dependent upon the availability of infrastructure which may or may not be in place. The early diffusion of the diesel car was limited by the lack of diesel pumps on garage forecourts. The distribution of pre-recorded video films relied upon the establishment of retail outlets. The diffusion of digital cellular telephones in place of analogue equipment will depend upon the provision of a total digital system. This can lead to a 'chicken and egg' situation. The infrastructure may not be provided until the products using it become available, but the product cannot be sold until the infrastructure is there. A judgment must be made about the rates of provision which both parties may be inclined to overestimate.

Customer implications

The customer can be affected in a variety of ways. He may need to acquire new skills in order to operate the product. Skilled servicing in a geographically dispersed location may replace local repair by relatively unskilled personnel. Environmental concerns may lead the customer to favour products which can eventually be recycled, a fact recognized in some new car designs. These factors may enhance or detract from the desirability in the eyes of the customer. If they are to have a significant influence on the acceptance of the product they must be identified at an early stage.

This outline is only an indication of some of the factors which should be assessed and will vary with the nature of the product and the industry. The important point is that it is essential to consider *all* the possible implications of a proposal, both internal and external to the company, before any significant funds are allocated to a project. Figure 4.1 illustrates a framework for a systematic examination. As a consequence it may be decided to reject the proposal at this stage, modify it so that it overcomes or minimizes some of the problems identified, or proceed to a detailed project evaluation.

Detailed evaluation

Whilst the first two stages are being undertaken the validity of the concept will

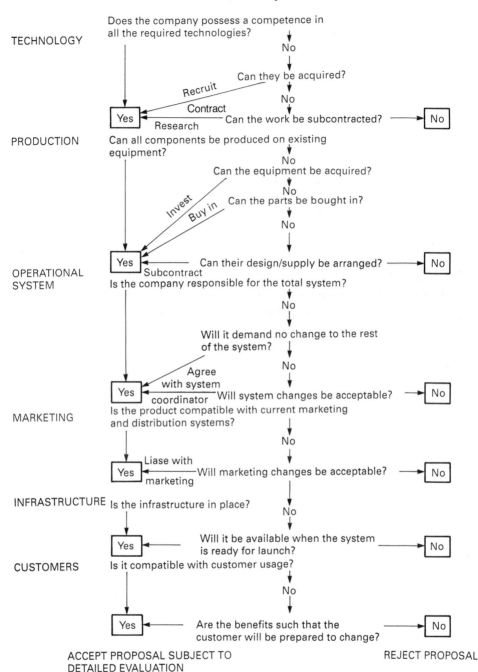

4.1 Project feasibility flowchart

have been established and the main features of the project clarified. It is now essential to carry out a detailed evaluation based upon a set of criteria and a methodology which enables the project to be evaluated in relation to them.

Selection criteria

The criteria which need to be considered in project evaluation differ with the circumstances of the individual company and its industry. Thus it is not possible to compile a comprehensive list for universal application. The factors discussed in this section, however, are likely to be relevant to most companies. It is, in any case, better to include some irrelevant factors that can be quickly discarded than to omit one which may at some time become critical for the commercial success of the project.

The cost of compiling a list of criteria is negligible. What can become costly is the gathering of information necessary for evaluating a project against the criteria. Judgment must be exercised in deciding which criteria are the most critical and the degree of accuracy in the data required for the purpose of decision-making. Resources are ill-spent when devoted to the refinement of information relating to a factor to which the viability of the project is relatively insensitive.

As we have seen already, many of these criteria are non-technological. Furthermore, it has been suggested that the early phases of R & D are devoted to the reduction of uncertainty. This uncertainty relates as much to the marketing, production and financial aspects of the project as to technological feasibility. Yet the term R & D is universally applied to the technological investment needed in order to establish that the projected innovation can be translated into a physical reality. It is customary to spend very little money outside the technical field until feasibility has been proved. Inevitably, the physical form of the final innovation has a major influence on many of the non-technological factors and limits the amount of useful information that can be gathered until there is some degree of certainty about the main characteristics of the product. Nevertheless, this is not entirely true and valuable information which could lead to the early termination of a project is often not gathered or used sufficiently soon because the company's organizational structure tends to confine evaluation of the project within the R & D department. The research studies referred to in Chapter 1 lead to the conclusion that many projects which later failed should never have been initiated, a situation which might have been avoided if due consideration had been given to a wider range of criteria in their early evaluation.

In summary, we conclude that the R & D phase should be concerned with the reduction of uncertainty in all areas which have a bearing on the commercial success of an innovation, that the selection and evaluation procedure should reflect all these factors, that the organizational structure should provide a framework within which this consideration can take place, and, finally, that it may be necessary to devote more resources than is customary to the collection of information outside the R & D department.

The implications of some of the most important qualitative criteria to be taken into account will now be discussed. For convenience they have been grouped together under the following headings:

1. Corporate objectives, strategy, policies, and values;
2. Marketing;
3. Research and development;
4. Financial; and
5. Production.

Corporate objectives, strategy, policies, and values

Strategic planning

The relationship between R & D decision-making and strategic planning for the whole business has been discussed in Chapter 2. The achievement of the corporate plan through the contribution of individual innovations depends in the final analysis upon the appropriate selection of projects. It is therefore vital that strategic considerations are reflected explicitly in the project selection procedure.

Corporate image

The concept of a corporate 'image' is closely linked to the strategic objectives. The image will have evolved as part of the character and value system of the company over many years. It is how the company and its products are perceived by its customers. This perception can usually be changed only over a period of years. Thus a new product inconsistent with this image may be rejected in the market-place irrespective of its intrinsic merits. A company, for example, which traditionally has produced cheap, mass produced, disposable goods may find it impossible to persuade the market to accept an expensive, high quality product because the potential customer will associate with it the standards he expects the company to provide.

Risk aversion

Corporate attitude to risk was discussed in Chapter 2. Selection of a high risk project particularly if it will subsequently involve a substantial investment of corporate funds, would be generally undesirable in a company where the top management has a high aversion to risk. This applies mainly to commercial risks since it is often possible to reduce technical risk by means of a limited R & D investigation.

Attitude to innovation

Top management's attitude towards innovation is closely related to their attitude to risk. Innovators are in general risk-takers also. Since the R & D director is the professional innovator within the organization, he may be expected to modify top management's attitudes towards innovation. The success of the project champion shows that this can be done but there are limits beyond which he is unlikely to succeed.

Three common fallacies in arguments against innovation are:

1. Comparison of the reliability of the new technology, usually poor, with that of the existing product or process. This ignores the considerable potential for improvement.
2. Consideration of the high initial cost without taking into account the fall with cumulative production volume due to the experience curve effect.
3. Satisfaction with the performance of the existing technology *vis-à-vis* the competition without assessing what that competition will develop in the future.

Time gearing

Time gearing is the emphasis which the company gives to either short- or long-term considerations. This is determined over a period by the corporate strategy, but is also influenced by the short-term economic climate. Although corporate objectives are clearly oriented towards the long term, it may still be necessary to subordinate these to more immediate problems when, for example, a major product is suffering from competitive pressures.

Corporate criteria are often less explicit and quantifiable than those that follow. It can be useful however, to consider profiles of risk, innovation and time gearing (Fig. 4.2) which, although subjective, can be helpful in revealing the likely attitude of top mangement to a project proposal.

Marketing criteria

Identifiable need

When the marketing department puts forward the initial suggestion for a new product, based either upon formal market research or the informal judgment of marketing management, it may be assumed that there is a high probability of success for a product which satisfies the need they have identified. There will, of course, still remain a number of unknowns which cannot be resolved until the new product assumes a sufficiently tangible form for the selling price/manufacturing cost/sales volume relationships to be estimated with some degree of accuracy. Moreover, during the period the new product is under development market needs may change or they may be satisfied by a competitive product or an innovation based upon a different, and perhaps superior, technological concept. Thus the existence of a clear need at the project selection stage does not mean the project's market future is sufficiently assured for this factor to be ignored in subsequent re-evaluations.

Project proposals resulting from new technology, particularly if they are radical, are much more difficult to relate to the market. Bright [3] has quoted radio, xerography and the computer to support his argument that radical

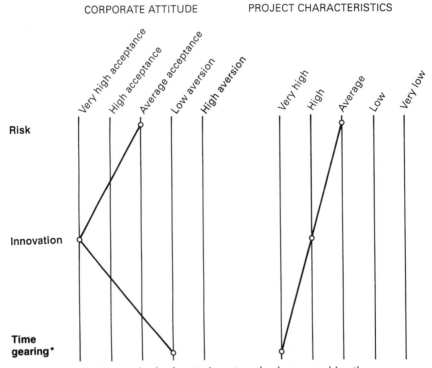

CORPORATE ATTITUDE   PROJECT CHARACTERISTICS

*Time-gearing is the emphasis given to long-term business considerations

Note: In this example the project is likely to be unacceptable on the grounds of
time-gearing, i.e. it is a long-term project whereas top management
attitudes favour results in the short term

4.2 Profiles of corporate attitude and project characteristics

technological innovations often achieve their first successes in fields not originally
envisaged. Furthermore, there are many cases where the first applications proved
costly failures. In Chapter 1 it was suggested that the technologist's enthusiasm
and his natural inclination to think in material terms are likely to encourage
him into embarking upon the development of a major application for a new
technology which may prove to have only a limited commercial potential. An
alternative conversion process, where the scientific or technological knowledge
is matched directly with customer needs or satisfactions before a product concept
is formulated, was advocated. If this matching can be achieved, it might be
expected to reveal the secondary applications where the unique capabilities or
the significant economic advantages of the new technology can best be exploited.
At a later stage, when the capability of the new technology has been
demonstrated, it can then be applied to other products where its technological
benefits are more marginal but its total potential market is greater. The choices
from which the initial application can be selected are illustrated in Fig. 4.3.

In Fig. 4.3, square 1 represents the technologist's dream which is very rarely

	Large operational economy or unique capability	Marginal economic or performance benefit
Large market potential	1	3
Small market potential	2	4

4.3 Innovation choices

realized (e.g. the microprocessor, float glass process). Most frequently the choice lies between an innovation in square 3 of which the technologist is aware, or searching for an initial application in square 2. Often it is preferable to seek a square 2 type solution for a radically new technology.

The application of the hovercraft or ground effect principle can be seen as an example of what so frequently happens. It was heralded as a major technological breakthrough which could revolutionize sea transport. It was seen in some quarters as a replacement for the wheel in a wide variety of uses. The major part of the development funds was devoted to building sea-going craft in direct competition with conventional ship designs and an alternative innovation – the hydrofoil. In the selected application it offered little operational advantage over conventional craft. To succeed it needed to demonstrate a significant economic advantage to reward early purchasers for the risks they were taking. With the benefit of hindsight one can see that the economic advantage was not significant, substantial unreliability was experienced, and the rate of adoption was so much slower than its developers had anticipated that several of the innovators experienced financial difficulties sometimes ending in bankruptcy. The availability of government development funds kept the project alive and helped to sustain it until the early operational problems were overcome.

In contrast, there are a number of applications, albeit of limited market size, where its technological attributes have satisfied a need in a way not possible using other technologies. These range from amphibious craft for patrolling swampy terrain, and the transport of heavy loads over weak bridges, to the design of special hospital beds to support patients suffering from widespread body burns.

The problem facing the innovator is to identify the attributes of the new technology and match them systematically with potential uses or needs to be satisfied. The possible applications do not follow automatically since, being new, they require the exercise of creativity. Nevertheless, the evaluation of the ideas in relation to the technology can and must be carried out methodically and objectively. Some of the technological attributes and possible applications for the hovercraft principle are shown in Fig. 4.4. The table has not been completed since the actual information to be used would depend to some extent upon the characteristics of the technical configuration.

POTENTIAL APPLICATION / TECHNOLOGICAL ATTRIBUTE	Sea ferry	Road transport	Amphibious patrol craft	Transport of heavy loads	etc.
Low friction					
Distributed load				High	
Atmospheric disturbance by exhaust					
Complete two-dimensional freedom of motion (for vehicle applications) etc.					
etc.					

Note: Each potential application is rated against all the attributes – for example, load distribution would be of high rating in respect of a transporter of heavy loads required to transverse weak surfaces (e.g. swamps, temporary bridges)

4.4 Matching technological attributes to potential uses − the hovercraft principle

Estimated sales volume

The volume of sales likely to be generated by a new product is one of the most difficult factors to estimate. The gross inaccuracy of many estimates used in project selection will be discussed in greater detail in Chapter 5. However great the problem, estimates do have to be made since sales volume is the final determinant of success. Clearly single-point estimates are inappropriate. Three-point estimates with associated probabilities for each of the parameters contributing to the estimate give a better understanding of the problem and its uncertainties. Although these estimates are also unlikely to possess any greater accuracy, they do at least give an indication of the margins of error and the sensitivity of the product's profitability to variations.

The technologist often experiences great difficulty in communicating any quantification of uncertainty to other members of the management team in meaningful terms. Three-point estimates and subjective probabilities may present the R & D manager with no conceptual problems. Nevertheless, other managers, while giving lip-service to their acceptance of uncertainty, will frequently press him for his 'best' estimate. His provisos and range estimates are quickly forgotten and these 'best' estimates will form the basis for their own decision-making, often with disastrous consequences. Similarly, the R & D manager may find the marketing department unused or unwilling to give him other than single-point estimates for the information he requires for his evaluation. This situation is improving as an increasing number of managers are becoming educated in quantitative techniques. But single-point estimates are still all too frequently used in evaluations under uncertainty.

The expected sales volume is derived from:

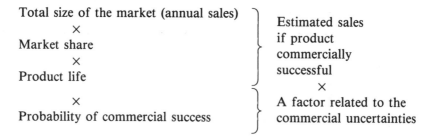

The accuracy of estimates for *market size* and *market share* vary with the nature of the new product. Market research is of assistance where the new product competes or replaces existing products in a well documented market familiar to the company. It is of less help and can be misleading when the product is entirely new. In these circumstances the response of potential customers to plans for an undeveloped and unpriced product can be a poor indication of their ultimate buying behaviour. Market surveys are thus likely to lead to highly optimistic forecasts when those whose opinions are sought do not have to back them with a decision to purchase.

Product life has a significant influence on profitability. In spite of the trend towards shorter product lives it is still possible to develop some new products with a life measured in decades rather than years. These 'bread-and-butter' products offer many advantages which make them attractive even when their profit margins may be lower than alternatives with shorter lives:

1. They assure the company of a steady income for an appreciable period.
2. They provide a continuing load for the manufacturing plant enabling specialized equipment to be depreciated over a high production volume.
3. They are less prone to fashion influences and in some cases provide better insulation from economic cycles.
4. Learning-curve effects and the scope for product improvement lead to reducing production costs.

Thus a number of factors difficult to quantify, particularly the advantages from stability and a steady income, favour the product which is expected to have a long life even when a simple economic analysis might indicate that it is marginally inferior to one with a shorter life. Financial discounting (DCF) can attribute a lower value to such products than management might wish to accord them.

The effect of product life on the utilization of manufacturing capacity is illustrated in Fig. 4.5, which compares two products having the same total sales volume. Product A, however, achieves its sales in only half the time it takes Product B and would, consequently, appear to be the better choice, since DCF calculations yield a higher net present value due to the earlier dates at which the earnings are achieved. Some economies of scale may be gained – sales forecasts for the shorter period may also be expected to be more accurate. On the other hand the manufacturing capacity required is twice as high. This might involve double the investment in specialized equipment utilized for what might be only a small proportion of its usable life.

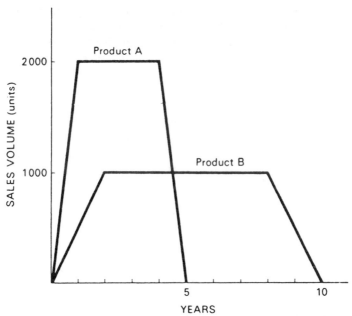

Note: Products A and B achieve identical sales volume during their product lives

4.5 Product lives and sales volume

The life of a product and its market share are vulnerable to competitive action. The threat from imitations or similar products marketed by a competitor may be as damaging as that from the introduction of an entirely new product. Thus it is necessary to assess the survivability of the product and what contributes to it. Amongst the factors to be considered are: the ability to patent features of its design and manufacturing processes, specialist skills in production and marketing, control of distribution channels and after-sales service.

A *probability of commercial success* multiplier is often used to modify a cost:benefit calculation in order to take account of the uncertainties inherent in estimates for both cost and benefit. No overall estimate for the probability of commercial success is necessary where each parameter in the sales volume calculation is associated with its own probability, usually subjective. The individual probabilities combine to give an overall figure of probability for the complete calculation. Success itself needs to be defined since sales volume is an insufficient criterion because of the price elasticity of demand. The business objective is to maximize earnings, i.e. the benefit in the cost-benefit calculation. Nevertheless, the motivation of marketing managers is often to maximize sales volume rather than earnings or profit and it is important for the R & D manager to establish clearly what is meant when a marketing manager speaks of 'commercial success'.

Where single-point estimates are made for market size, share, etc., it is useful to apply an overall subjective probability for commercial success. Nevertheless, this is less desirable than a series of estimates for the constituent parameters where the mind is focused on making individual judgments in respect of a number of items. The relative importance of each item is thereby identified and attention directed to those where further data should be sought.

Timescale and relationship to the market plan

The launch of a new product requires a considerable investment of the marketing department's managerial and financial resources. They too have portfolio considerations to take into account and their ideal would be a regular succession of new products emerging from R & D, sufficient to replace those they wish to withdraw from the market and to meet planned expansion of the product line. Unfortunately, successful new products cannot be developed to order so as to arrive in the right numbers at the time dictated by marketing requirements. Nevertheless, attention needs to be paid to this aspect if bunching of new products is to be avoided, thereby exceeding the ability to introduce them on to the market in an orderly fashion. This is particularly important where product lives are short and the timing of market launch is critical, or where the new product is of sufficient importance to warrant the investment of a major portion of the marketing department's effort.

Although it is unknown whether a new project will result in a saleable product when it is selected, some control over the rate and timing of new products can be maintained by ensuring the right balance within the R & D portfolio. As development proceeds increasingly accurate estimates can be made of the date when it is likely to be ready for market launch. Very occasionally a company may be fortunate in having several new products likely to emerge within a short time, in excess of what the marketing department can handle. At such times it may be necessary to terminate or temporarily shelve a project which at other times would have been supported. On another occasion it may be necessary to continue development of less desirable new products to meet a serious deficiency in the product line.

Effects upon current products

New products may be additions to the product line or partial or complete replacements for existing products. In the latter case a proportion of the sales volume (and profit contributions) will be achieved at the expense of the existing product which may still be selling successfully. An entirely new product extends the range and its sales volume and profits contribute to corporate growth.

Projects leading to entirely new products appear, therefore, to be preferable to those whose sales will only partly contribute to corporate growth. But it is important that this undoubted advantage should not lead to the total neglect of the current product range. A successful sales record may easily lead to complacency. The success of a product contains the seeds of its own downfall since this is a visible challenge to other companies, who may introduce their own competitive products possessing new features giving them a market advantage. It is known that the current product satisfies a market need, has established customer confidence and represents a valuable company investment; this position of market strength must be safeguarded by product improvement and its occasional replacement by a new product to maintain the competitive edge. A decision to give priority to the development of entirely new products, attractive though this may be to the R & D department, is bound to be associated with a greater degree of uncertainty than support for the established product, whose continued development should only be abandoned if it is nearing the end of its useful life, or there are other reasons for wishing to withdraw from a particular market.

Thus an assumption that a project leading to a product competitive with current products is undesirable may not be correct in certain circumstances. Although it is important that this factor be included in the evaluation, judgment of whether it weighs in favour of the proposal depends as much upon careful consideration of the current product, as upon the benefits expected from an addition to the product range.

Pricing

In discussing the expected sales volume the price elasticity of demand was not considered. Sales volume representing customer acceptance is, however, a function of the price charged, which in itself depends upon the perceived value of the product to the potential market as well as the pricing of competitive products. A full discussion of pricing will be found in the marketing literature [4, 5]. For our purposes it is sufficient to recognize that the price of a product should be determined by market forces whereas product cost is within the control of the manufacturing company. The profit which is sought from the new product is a residual, highly sensitive to changes in both price and cost, difficult to assess at a time when the final form of the product has not been established. Optimism is, however, a characteristic of project assessments. Thus, as development proceeds estimates of the price at which the product can be sold are likely to fall accompanied by rising estimates for manufacturing cost (Fig. 4.6).

In Fig. 4.6 hypothetical estimates during the development period for two projects are plotted. At the time of project initiation Project A appeared to possess a better potential profitability than Project B whereas this position has

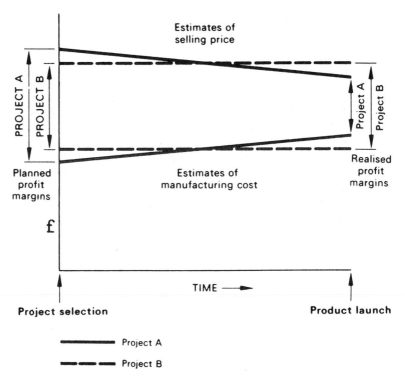

4.6 Price and cost estimates during project development

been reversed by the time the products are ready for launch. It is of course impossible to foresee how the estimates will change, for the initial ones were, after all, the best that could be made at that time. Nevertheless, in project selection it should be recognized that:

1. Price estimates are almost invariably optimistic. Consequently projects which only meet the company's investment criteria by a small margin on the basis of early estimates are unlikely to achieve them in practice.
2. Estimating errors may change the relative attractiveness of projects as development proceeds. The more radical the innovation, the more likely are the estimates to change significantly (i.e. Project A in Fig 4.6 is likely to be more radical than Project B). Thus greater confidence can be expected

in estimates for projects leading to product improvements or the replacement of existing products than for entirely new products.

Competition

The importance of assessing one's strengths in relation to competitors has been stressed in the formulation of corporate and R & D strategy. It was seen that care needs to be taken to identify who the competition really are, since technological change frequently alters it. Threats to established products may come from 'innovation by invasion' by companies in industries which have not traditionally been regarded as competitors.

In assessing a project which will not emerge as a product for several years it is easy to underrate the pace of a competitor's own technological advance. It is not sufficient to develop a product better than his existing product. Comparison must be made with what he might reasonably be expected to have on the market at the time one's own new product is introduced.

Distribution channels

A new product will normally be capable of distribution through the existing logistic system unless it appeals to an entirely new market. If new distribution channels do have to be established the cost can be high, and may well tip the balance of profitability against a new product proposal. This is one of those factors which are frequently ignored until late in the development cycle, yet the situation is unlikely to change very much during this period.

Launching costs

The marketing department can usually estimate launching costs for a new product with reasonable accuracy, although their magnitude is dependent upon the type of product and nature of the market. An industrial product advertised mainly in the technical press is purchased by well informed customers able to analyse the technical merits of the product for themselves. This is likely to be considerably less expensive than the advertising in press and TV necessary for launching a new consumer product.

The nature of the product also has a considerable influence upon the marketing investment necessary to launch it. New customers must be persuaded to buy the product. New customers, and those who have previously bought the company's products, have to be convinced of the merits of a new product particularly if it introduces a radical technological innovation. All this costs money, estimates for which must be included at every stage of the project's evaluation.

Research and development criteria

So far our discussion has centred upon the desirability of the product or process resulting from the application of R & D resources. It is now necessary to turn our attention to the desirability of the project itself, first in respect to the likelihood of it achieving the technological performance required and secondly, the effect it will have upon a laboratory with limited resources.

In spite of the need to stress the importance of relating R & D activity to the requirements of the business as a whole, it must not be overlooked that the R & D department is itself an organization which can continue to contribute to the long-term prosperity of the company only so long as its own internal health is maintained. In this its needs are neither greater nor less than those of other parts of the business. The undertaking of projects which sap the R & D department's strength is undesirable, however attractive they may seem from other points of view. Thus, when considering a new project in relation to R & D criteria its effect upon the total activity of the laboratory, present and future, must be borne in mind as well as the likelihood of its technical success.

Consistency with the R & D strategy

The development of the R & D strategy has been considered in detail in Chapter 2. The strategy is implemented through the selection of projects which provide a balanced portfolio designed to further corporate objectives while at the same time ensuring the optimum use of departmental resources over a range of timescales. Nevertheless, it was seen that the process of technological innovation cannot be constrained within rigid limits and projects which do not fit the strategy should not be rejected automatically.

Probability of technical success

Although technical success is often defined as the achievement of the technical performance required in the project specification, it is more useful to add that it must be obtained within the budgeted cost and the required timescale. In most industries, technical success is a function of the financial resources available. Research directors are unlikely to support projects they are not confident are technically feasible. They may be much less confident of which of several technical approaches might be the best alternative for achieving their specifications. Nor can they foresee all the problems that will have to be overcome.

When there is some doubt regarding the attainment of a particular aspect of a project it is not uncommon to follow parallel approaches using, perhaps, different technologies or to precede the main project with an investigation into the aspects of the problem where major technical uncertainties exist. For example, the development of a military ballistic missile was thought to be open to question on two counts − the feasibility of designing a capsule which could re-enter the earth's atmosphere without burning up, and the development of a sufficiently accurate inertial guidance system. The re-entry problem was tackled by the development of a test vehicle with a variety of nose cones to establish whether warheads could survive the re-entry environment. The inertial guidance system was supplemented by the parallel development of a radar guidance system, subsequently abandoned when inertial guidance development had reached a stage where its feasibility was established. In both cases the initial fears proved groundless.

Another example is the development of carbon fibre Hyfil blades for the Rolls-Royce RB211 aero-engine. When these blades encountered development difficulties the production programme was switched to titanium blades, development work on which had been undertaken as an insurance. The discontinuance of the Hyfil development resulted from the cost of solving the problem and the demands imposed by the timescale for the development of the engine rather than lack of confidence that the trouble was solvable given time and money.

The fact that it is rare for a project to fail to achieve its technical objectives may reflect a conservative attitude of R & D directors or a tendency to underestimate the pace of technological advance. Where serious doubts exist, a parallel approach usually ensures that at least one feasible solution is obtained. In project selection, therefore, it can be assumed that the probability of technical success is high for any proposal that survives its initial screening. The relevant considerations are not whether the project is feasible but how much investment is necessary and when the development will be completed. A possible exception to this generalization is found in the chemical and pharmaceutical industries.

Development cost and time

Estimates for development cost and time to complete the project are amongst the important inputs to the evaluation system and will be discussed in greater detail in Chapter 5. Apart from their importance in assessing the financial viability of the project they provide a measure of the scale of R & D resources absorbed by the project and its duration. To that extent they have an effect upon every other project being undertaken in the R & D department.

Patent position

Little need be said about the patent position other than to stress that, before major resources are allocated to a project, it is essential to ensure that a patent held by another manufacturer is not being infringed. Furthermore, the ability to patent one's own developments can safeguard the product's long-term competitive position.

Availability of R & D resources

The research director must find the money to pay for a new project from his annual financial budget. Unless the project is of exceptional importance, enabling him to obtain additional funds, it must be accommodated within his existing budget.

The financial budget is, however, only a convenient measure for aggregating expenditures on a wide variety of resources which are not interchangeable. A physicist, for example, cannot do the work of a chemist, nor does surplus capacity on an electron microscope alleviate over-usage of a fatigue-testing machine. Thus, in considering the addition of a new project to the portfolio, it must not only be evaluated in relation to its total budgetary impact but also its demand for other scarce resources. To do this, estimates must be projected forward for several years in much the same say as the financial director produces his cash flow projections (Table 4.1).

Table 4.1 *Availability and utilization of manpower – physicists*

| | Man years | | | | | | | | | | | |
| | 1993 | | | | 1994 | | | | 1995 | | | |
Project	Jan–Mar	Apr–Jun	Jul–Sep	Oct–Dec	Jan–Mar	Apr–Jun	Jul–Sep	Oct–Dec	Jan–Mar	Apr–Jun	Jul–Sep	Oct–Dec
1	0.25	0.25	0.25	0.50	0.50	0.50	1.00	1.00	0.50	0.25	–	–
2	1.00	2.00	2.50	2.50	1.00	–	–	–	–	–	–	–
3	–	–	–	0.50	0.50	1.50	1.50	2.00	2.00	1.00	0.50	–
4	2.00	1.50	1.50	–	–	–	–	–	–	–	–	–
5	0.25	0.25	0.25	0.25	0.25	0.25	0.25	0.25	0.25	0.25	–	–
6	–	–	0.50	0.50	0.50	0.50	1.00	1.50	1.50	0.50	–	–
New Project X	–	–	–	0.25	1.00	1.50	2.50	2.75	2.75	3.50	3.50	2.00
Total requirement	3.50	4.00	5.00	4.50	3.75	4.25	6.25	7.50	7.00	5.50	4.00	2.00
Total available	4.00	4.00	5.00	5.00	5.00	5.00	5.00	5.00	5.00	5.00	5.00	5.00

From a series of such projections the research director is able to assess the impact of the new project not only in the immediate future but throughout its development. Although a high degree of accuracy cannot be expected from these estimates they do enable the identification of major areas of mismatch between availability and utilization; minor difficulties can be regarded as routine scheduling problems. Table 4.1 suggests that the addition of project X to the portfolio creates a shortage of physicists unlikely to be satisfied completely by minor adjustments to the work schedule. In this situation the research director is faced with several alternatives:

1. To reschedule the allocation of physicists to other projects, thereby affecting their planned completion dates.
2. To reschedule project X.
3. To recruit additional physicist (1) in July 1994.
4. To terminate one or more of projects 1 to 6.
5. To abandon project X.

Ideally, the R & D director would like to have a portfolio of projects progressing at their own optimum pace and completely utilizing all his resources the whole time. In practice, he must fit together his project programme by changes to his resources, scheduling of projects or modifications of the portfolio itself. Thus, a shortage of a particular resource may become the deciding factor in choosing between projects which appear equally attractive in most other respects.

Future development

A product innovation may be an isolated development or the first of a family of new products providing the company with a low risk 'bread and butter' product line. Similarly, the knowledge or technological capital generated by some R & D programmes is unique to a particular project, whereas in other cases it can be relevant to a variety of future projects. Thus evaluation of a project should extend beyond the immediate new product and give some recognition to the future benefit to be derived from the security afforded by a family of products developed over a period of years, and the use of its technology either to further developments of the product or in other applications. The farther one looks into the future the more difficult it is to put a monetary value upon these benefits. Nevertheless they exist, and the desirability of continuity in product range or technical effort should be reflected in the financial criteria applied when comparing such projects with others of more limited potential.

A word of warning should, however, be sounded. Future benefits ought not to be used to justify a project which would otherwise be commercially unsound. Technological 'spin-off', although frequently of great value, is often unquantifiable and difficult to foresee precisely. If the future potential is worth acquiring, then it ought to be justified and funded on its own merits and not on the back of a supposedly commercial project. For the combination of two marginal requirements rarely leads to the satisfactory conclusion of either. Nevertheless, the synergistic effect is to be encouraged provided that judgment is not clouded.

Environmental effects

Public concern about the environmental and ecological effects of technology is becoming a factor of increasing importance in industrial decision-making. Many products or processes acceptable today are under increasing attack from environmentalists; some will be banned as a result of legislation, others, while remaining within the strict interpretation of the law, will harm the company's reputation if left on the market. These pressures have already affected a number of products – detergents containing phosphates, DDT, aerosols. Frequently the technologist will disregard the environmentalists because he 'knows' their arguments are factually incorrect. But the 'fact' he must be aware of is that the pressures are real whether he likes it or not or whether or not he considers them unsoundly based. Such decisions are difficult to make since most products contain some degree of inherent risk. For example, one manufacturer decided not to market a new cosmetic product because of a slight doubt regarding possible carcinogenic effects whereas a competitor assessing the risks differently launched a similar product successfully.

Product-liability legislation is growing, particularly in the USA. This is a consideration which must be taken into account early in the life of a project. Faulty products (e.g. car safety, tyres, etc.) can involve the manufacturer in expensive modification programmes and litigation.

It is becoming important, therefore, that the possible environmental effects of a new project should be considered carefully at the selection stage. In this, social and political forecasting may be necessary to assist in anticipating what the new product's environmental status will be throughout its expected life. Some companies are already doing this. Dow Chemical Company for example has established an 'Ecological Council' and research groups are required to assess in advance their products' impact on the environment. A British chemical manufacturer terminated a promising research project because it was doubtful whether the product would be acceptable in the future, although it met fully current regulations.

Financial criteria

In considering the financial aspects of a new project it is important to draw a clear distinction between its ultimate profitability and the demands it places upon the company's financial resources before it commences to make a financial contribution. Evaluation of project profitability (see Chapter 5) normally receives a great deal of attention. A highly favourable estimate of the eventual profitability of the project may distract attention from the drain upon the company's financial resources required for its development particularly in the likely event of cost escalation. This is of greatest significance where one or two projects are very large in relation to the company's total resources. Cash shortages brought Rolls-Royce to bankruptcy in 1973 due to the escalating development costs of the RB211 engine. That engine and its derivatives then became the foundations for the subsequent growth and profitability of the company under new ownership, likely to extend well into the twenty-first century. In this section the cash flow aspects of a project will be considered, discussion of profitability being left to the next chapter.

Cash flow

The main expenditures incurred before a new product or process begins to earn a return on its investment consist of R & D costs including prototype or pilot plant construction, capital investment in manufacturing plant, and initial marketing costs. The relative importance of these items depends upon the nature of the industry. In aerospace, for example, the major expense is usually the construction and testing of prototypes and the facilities required for doing so, whereas in the chemical industry the manufacturing plant may be the major item, and in consumer industries the marketing costs in launching the new product.

The curve in Fig. 4.7 illustrates the cumulative investment required in a project to the time of financial breakeven. Although the shape of the curve and the relative magnitudes of the investment needed at the various stages will vary with the product and the industry, the rapid increase in costs as the product approaches market launch is typical. The financial director is concerned with both the magnitude of the costs involved and their timing since, assuming the ultimate financial viability of the project is not in doubt, his problem is to ensure

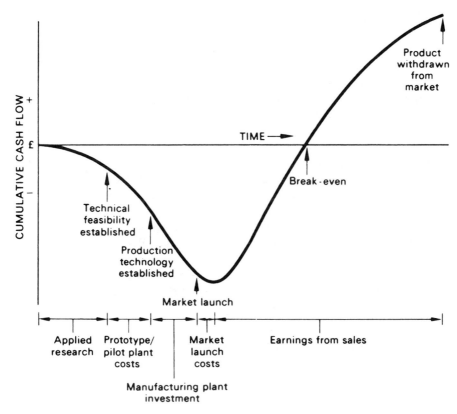

4.7 Project cumulative cash flow diagram

that the finance is available when required. Unfortunately, there are few projects where either the total costs or the timing can be estimated with any degree of certainty at the initiation. The availability of the finance is dependent upon the cash flow position resulting from all the activities of the company, both inflows and outflows. Thus finance availability is highly sensitive to timing. Few companies seem aware of the shape of their cash flow profile. Readers are urged to draw one for a typical project. They are likely to be surprised by the rapidity with which the cash flow mounts as a development progresses.

Where financial availability is likely to prove a critical factor it is essential that the cash flows involved are estimated as accurately as possible and their implications fully understood. This analysis may show:

1. The maximum negative cash flow is well within the capacity of the company's finances. In these circumstances, cash flow limitations are unlikely to impose a constraint in project selection.
2. The cash flow requirements may be marginally within the expected availability of capital. This might, however, make the company vulnerable to an escalation of development costs or a deterioration in cash inflow due to

adverse trading conditions. These risks must be recognized and steps taken to avoid them in good time by:

(a) rephasing the development programme to spread the expenditure over a longer period, or to plan for the maximum investment to coincide with a time when the overall cash flow is temporarily favourable; or

(b) making contingency plans to raise the necessary finance before a cash shortage becomes critical.

3. The requirement for funds may exceed the likelihood of their availability. In this situation the project, however desirable it appears in all other respects, might be dropped. There are, however, other alternatives which should be considered such as partial development and then licensing or a joint venture with another manufacturer. In exceptional circumstances the project potential may enhance the worth of the company sufficiently to make it attractive for takeover. None of these solutions should be ignored provided they are accepted and pursued as a conscious act of policy. Perhaps few companies are likely to invest in a technological innovation with the avowed objective of it leading to the taking over of the firm. Nevertheless, it frequently happens, usually when faced with a cash crisis which seriously undermines the company's bargaining power.

Effect upon other projects requiring finance

The financial director, like the R & D and marketing directors, is also concerned with a portfolio of investment opportunities. These embrace the whole gamut of company activities including diversification, takeovers, extension, and modernization of manufacturing facilities as well as the financing of projects emerging from the R & D department. His resources are limited and consequently no investment opportunity can be examined in isolation. Short-term financial as well as long-term strategic considerations have to be taken into account. Thus occasions may occur when projects placing a heavy demand upon capital resources should be rejected in favour of less profitable but less capital-consuming projects, not because there is no cash available but because it is required for investment elsewhere. Such decisions are bitterly resented by R & D directors, nevertheless these factors must be taken into consideration, for there is little logic in devoting scarce resources to projects which can never be developed fully because the financial resources they would absorb are needed elsewhere in the business.

Production criteria

The introduction of a new product into the manufacturing organization is rarely accomplished without difficulty. Many of the problems which are likely to occur can be foreseen while the project is still within the R & D department; some of them are sufficiently important to be considered at the earliest stage of project evaluation. These difficulties may be broadly categorized under two headings:

1. The capability to manufacture the new product.
2. The ability to manufacture at a cost which will leave an adequate profit margin.

Manufacturing capability

When discussing the development of a strategy it was seen that a major consideration was the capability of an organization to exploit potential opportunities. This capability stemmed from the strengths or weaknesses of the organization. A new product can be likened to a new opportunity for a manufacturing department and its ability to make a success of it depends upon the accumulated experience in manufacturing similar products, the number and skills of its personnel and the suitability of existing production equipment.

By the time a project nears completion and moves from R & D to manufacturing the feasibility of producing small quantities of the final product using specialized equipment and highly skilled personnel will have been established. But the speed and cost of setting up large-scale production varies considerably from product to product. Delays due to teething troubles, the purchase of new equipment, and the recruitment or training of production personnel in new skills, all have a financial impact which should be taken into account in project evaluation even though they cannot be quantified accurately. The greater the mismatch between the needs of the new product and the existing capability of the manufacturing department the greater this cost will be.

In project evaluation it is important to identify those characteristics of the new product which are likely to pose manufacturing problems, not so much in the laboratory or in an ideal production organization, as in the actual plant ultimately responsible for it. The costs involved and the relationship to the company's manufacturing policy are pertinent matters for consideration before embarking upon the project, even though detailed planning may be some time away. Furthermore, even at this early stage it may become evident that some effort may need to be devoted to investigating anticipated production problems, a matter often deferred until so late in the development cycle that there is inadequate time to carry out in sufficient detail all the preparations necessary for the smooth transition from R & D to production.

Cost of manufacture

The final cost of the product depends upon the price of raw material and purchased parts, the manufacturing technology, the labour man-hours and skills, the capital investment and the volume of production. Although the ability to manufacture is rarely in doubt it is by no means certain that the final product cost can be kept sufficiently low to achieve an adequate return on the investment. In many cases the interrelationship between (a) the manufacturing technology and cost, and (b) the sales volume and price can be critical for success. These interrelationships are sometimes highly complex (Fig. 4.8).

Figure 4.8 shows the price and cost relationships for a 'Product X' which

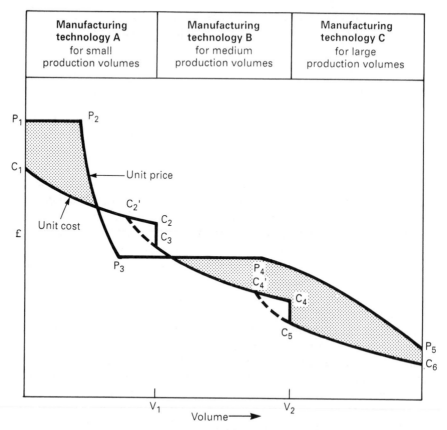

Manufacturing technology A for small production volumes	Manufacturing technology B for medium production volumes	Manufacturing technology C for large production volumes

$P_1 - P_2 - P_3 - P_4 - P_5$ represents the price at which the corresponding volumes can be sold

$C_1 - C_2 - C_3 - C_4 - C_5 - C_6$ represents the manufacturing costs at the corresponding production volumes

The shaded portion represents regions of profitable operation

4.8 Interrelationship between manufacturing technology, plant size and price – Product X

can be manufactured in three different ways, according to the expected sales volume. Analysing the two curves we observe the following:

Price

$P_1 - P_2$ A high price can be commanded for a specialized market which cannot be satisfied by any other product.

$P_3 - P_4$ In this price range Product X is competitive with an existing product where substitution occurs because of a price advantage once user doubts concerning reliability and suitability have been dispelled.

$P_4 - P_5$ Further growth by substitution in other applications as the price falls.

Cost

C_1-C_2 Falling manufacturing cost with Technology A due to economies of size and utilization.

C_2-C_3 Step drop in manufacturing cost when volume V_1 reaches economical size for Technology B. *Note* although Technology B becomes competitive at C_2 the decision to move to the larger plant is usually delayed until some time later.

C_3-C_4 Falling manufacturing cost with Technology B due to economies of size and utilization.

C_4-C_5 Step drop in manufacturing cost when Volume V_2 reaches economical size for Technology C.

C_5-C_6 Falling manufacturing cost with Technology C due to economies of size and utilization.

In this illustration Product X is only profitable in the shaded portions of the graph, even so this is only within the limits of accuracy of the estimates. Achievement of these profit levels also presupposes that market penetration is rapid – in practice there may be a considerable time-lag between the setting of a price and the realization of the expected sales volume, since the inherent uncertainties with a new product may outweigh the price advantage in the minds of many potential customers.

A great deal of analysis is necessary for a clear picture of the alternatives to emerge and the expense may only be justified for major projects. Furthermore, accurate data is unlikely to be available before the project is well advanced. Nevertheless, the considerations which give rise to the differences in cost and profitability do have an important bearing on the form the project takes, for it is unlikely that the products for the different markets will be identical. Thus, what starts as one project may develop into several alternative projects leading to products suitable for different markets and manufacturing technologies. The sooner this dilemma can be resolved the better. Frequently this will lead to a conscious decision for a phased development of the project with the initial effort concentrated on satisfying the high-margin low-volume market.

Value added in production

Different products make varying demands upon production resources. Some require extensive metal forming operations utilizing a high proportion of the company's manufacturing capacity, while others consist largely of the assembly of purchased components. All manufacturing organizations are faced with the problem of maintaining a high rate of capacity utilization since the unabsorbed overhead cost of low utilization has a marked effect upon company profitability. This problem is illustrated in Fig. 4.9 which shows a company's productive capacity and its planned utilization by existing and new products. At some periods there will be spare capacity through the phasing out of old products (T_1-T_2) or factory extensions (T_4-T_5); at other times $(T_3$ and $T_5)$ planned production will approach or exceed capacity. Ideally, new products added during

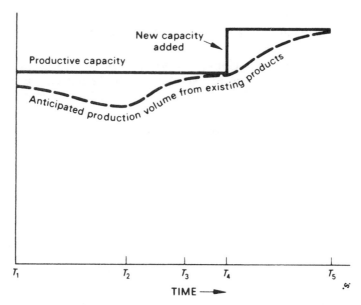

4.9 Utilization of productive capacity

the periods T_1-T_2 and T_4-T_5 should have a high value added in production, i.e. the manufacturing costs are a high proportion of the total costs and conversely during T_3-T_4.

It is unrealistic to suggest that projects can be selected with a view to matching accurately the development of new products to the future availability of productive capacity. However, there may be occasions when forecasts of future demand and new product availability do reveal the likelihood of major under- or over-utilization of capacity. Reorientation of the R & D programme can be effected to reduce under-capacity by emphasizing process developments to improve productivity or new products with a low added value in production or, alternatively, to utilize spare capacity by developing new products with a high added value in production.

Discussion

A large number of factors have now been examined which affect the eventual commercial viability of a technological innovation. The reader may be tempted to think that the examination of such a wide range of criteria makes the selection and evaluation of projects a matter of such complexity, in view of the major uncertainties involved, that it is impossible to approach the problem systematically and that any formal evaluation system is unlikely to be meaningful. It should be equally obvious that straightforward economic analysis is an over-simplification since many important non-quantifiable factors cannot be included in the calculation.

It is of course true that many of the criteria are of limited importance in mos† evaluation decisions and that many of them are often impossible to assess at the time of project selection. Yet each one of them may be of particular significance at some point of time and should not be lost sight of. The accumulating research evidence shows conclusively that failure frequently results from lack of consideration of some factor which is critically important for the success of an innovation. Often it is the ultimate market potential of the innovation that is disregarded. The survey *On the Shelf* [5] reports:

> It is wasteful to develop a product for which there is no demand and yet the survey suggests that this happens not infrequently. Market information was not obtained or was inadequate in several of the cases, as a result, work was commenced which should not have been or was directed at obtaining the 'wrong' characteristics for market requirements.

and:

> Finally firms may inadequately study the competitive environment. The Data Generator was shelved after it was established that rivals were already producing an adequate instrument. Had this been noted earlier, given the company's policy, the R & D effort might not have been initiated.

Who should evaluate?

The majority of the criteria considered have been non-technological. The influence of an innovation, be it success or failure, spreads throughout the company, and in the final analysis it becomes an integral part of the business. It affects not just one department in isolation but several elements of the dynamic system called the business. In most companies the selection of R & D projects is vested in the R & D director, other functional directors and top management only being involved at this stage if the project is likely to be of particular significance to the company or involve a substantial initial investment. Several R & D directors have strongly supported this procedure in discussion with the author, maintaining that project selection is a technological matter, of departmental concern, which should only involve other departmental managers when the project has been proved technically feasible.

This attitude can be understood but not necessarily accepted. While it is understandable that the R & D director should wish to retain control of his resources, many of his decisions cannot be taken without giving due consideration to factors arising outside his department. This cannot be done effectively without the involvement of those concerned. If it is accepted that project evaluation should involve a wide functional representation it must be introduced in a way which does not appear to reduce the status of the R & D director. Thus, as much attention needs to be paid to the process of project evaluation as to the particular technique to be used. This process must encourage the meaningful cooperation of all those managers who can contribute information necessary for an evaluation to the depth appropriate to the decision or who will later be responsible for implementation of a stage in the innovation chain. In Chapter 7 organizational structures and their influence on interdepartmental relationships are examined. The structures provide a framework within which specialist functional areas interrelate. Procedures such as project evaluation contribute to the operational cooperation within the structure. However, it must be stressed that although the choice of an organizational structure and operational procedures is important, as much attention needs to be paid to the behavioural processes and the maintenance of good interpersonal and intergroup relationships. If these do not exist, a formal evaluation scheme becomes meaningless; it may result in yet another forum for interdepartmental conflict. Fears that this may occur explain why many R & D directors adopt an isolationist approach and resist the involvement of other departmental managers in evaluation decisions. Differences in emphasis will inevitably occur but every effort must be made to achieve consensus.

Thus evaluation is more than a simple procedure. Where it operates effectively it can become an important element in the innovation process enabling the identification of managers outside the R & D department with the project at an early stage. This identification leads to commitment and eases the transition of the project from R & D to production and finally to marketing.

Evaluation techniques

The simplest form of evaluation technique is to assemble a *checklist* (Table 4.2) of all the criteria which need to be taken into account when evaluating the project. This ensures that none is disregarded even though it may be difficult to evaluate many of them when the project is initially selected. Research studies show that checklists are widely used but it is surprisin₁ how few companies attempt to compile a comprehensive list as a basis for formal evaluation.

Table 4.2   *Checklist of project evaluation criteria*

A. *Corporate objectives, strategy, policies, and values*
1. Is it compatible with the company's current strategy and long-range plan?
2. Is its potential such that a change in the current strategy is warranted?
3. Is it consistent with the company 'image?'
4. Is it consistent with the corporate attitude to risk?
5. Is it consistent with the corporate attitude to innovation?
6. Does it meet the corporate needs for time-gearing?

B. *Marketing criteria*
1. Does it meet a clearly defined market need?
2. Estimated total market size.
3. Estimated market share.
4. Estimated product life.
5. Probability of commercial success.
6. Likely sales volume (based on items 2 to 5).
7. Timescale and relationship to the market plan.
8. Effect upon current products.

9. Pricing and customer acceptance.
10. Competitive position.
11. Compatibility with existing distribution channels.
12. Estimated launching costs.
13. Vulnerability to competitive responses.
14. Compatibility with the existing infrastructure.

C. *Research and development criteria*
1. Is it consistent with the company's R & D strategy?
2. Does its potential warrant a change to the R & D strategy?
3. Probability of technical success.
4. Development cost and time.
5. Patent position.
6. Availability of R & D resources.
7. Possible future developments of the product and future applications of the new technology generated.
8. Effect upon other projects.
9. Compatibility with the total operational system.
10. Availability of software.

D. *Financial criteria*
1. Research and development cost:
   (a) capital
   (b) revenue.
2. Manufacturing investment.
3. Marketing investment.
4. Availability of finance related to timescale.
5. Effect upon other projects requiring finance.
6. Time to breakeven and maximum negative cash flow.
7. Potential annual benefit and timescale.
8. Expected profit margin.
9. Does it meet the company's investment criteria?

E. *Production criteria*
1. New processes involved.
2. Availability of manufacturing personnel − numbers and skills.
3. Compatibility with existing capability.
4. Cost and availability of raw material.
5. Cost of manufacture.
6. Requirements for additional facilities.
7. Manufacturing safety.
8. Value added in production.

F. *Environmental and ecological criteria*
1. Possible hazards − product and production process.
2. Sensitivity to public opinion.
3. Current and projected legislation.
4. Effect upon employment.
5. Recycling potential.

Note:
This checklist is not comprehensive and suitable for universal application. Nevertheless most companies are likely to find the items listed relevant in project evaluation.

A checklist compiled by the Industrial Research Institute is divided into three categories − research; product development and process development (Becker [5]).

It might be argued that the studies are not truly representative since some additional factors may be taken into account although excluded from the formal evaluation procedure. Nevertheless, the conclusion seems clear: few companies use comprehensive checklists. As a result some important business considerations may be ignored and this could be one explanation why resources are so often devoted to projects which fail to lead to successful innovations.

A development of the checklist is the project profile where each of the criteria is evaluated against a standard of performance (Fig. 4.10). In completing the

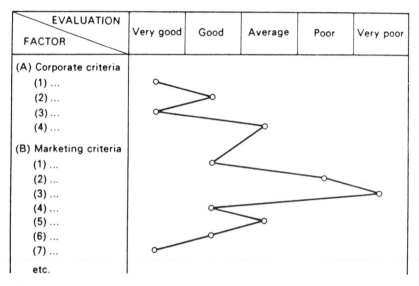

4.10 Project profile

profile a judgment must be exercised for each item based upon quantitative analysis where possible and compared with a formal or informal understanding of what constitutes 'very good', 'good', etc. Having reached this stage, some companies using project profiles have been tempted to add the number of items in each column to give an overall quantitative assessment of the project. This can be misleading because:

1. Some criteria are of overriding importance (e.g. a 'very poor' in respect of patent position should normally lead to rejection of the project irrespective of its other merits).
2. Some criteria are more important for the success of the project than others; this is not reflected in a profile where all items are weighted equally.

A further development is to devise a *merit number* system, in which the individual items are given a weighting in accordance with their importance and each evaluation grade is also given a value (Table 4.3). The calculation results

Table 4.3 *Merit number calculation*

Factor	Factor importance weighting	Evaluation against standard					Item merit value
		Very good 5	*Good* 4	*Average* 3	*Poor* 2	*Very poor* 1	
A. *Corporate criteria*							
1	10	5					50
2	8		4				32
3	8				2		16
4	9		4				36
B. *Marketing criteria*							
1	10			3			30
2	6					1	6
etc.							

Total – Overall merit number 627

in a merit number which purports to give an indication of the 'value' of the project. Although a merit number leads to a simple quantitative criterion for selection, one may question how useful an indication it gives of the real worth of a project. So many subjective judgments are involved in deciding the weightings, the assessment of the project in respect of each factor, and the arbitrary allocation of a numerical value to each grade (i.e. 'very good' 5; 'very poor' 1), that the system is probably of little real value although its supporters argue that it does focus the mind on these considerations provided the final merit number is interpreted with care.

There are many other evaluation techniques which are used but most of them focus on only a few of the quantitative factors [8]. Some of these are discussed in Chapter 5. No overall evaluation system can combine simplicity with accuracy. Nor can the procedure remove the uncertainties inherent in the information or absolve those operating it from using their judgment, a judgment inevitably coloured by their personal attitudes to the project. In short, no technique can do the manager's job for him, yet he should make use of any procedure which he finds personally useful to him in forming his judgment. A comprehensive checklist appears essential but whether any sophisticated elaboration of it is of real help must be questioned. He must also guard against the temptation to 'doctor' his evaluation to give the result he wishes through the use of spurious quantification.

It must be stressed that the use of any evaluation system should not be mechanistic. Unfavourable factors are not absolute since they relate to the current situation. Their identification should spur consideration of ways in which they can be overcome. For example, an inability to produce or market might lead to an investigation of alternatives such as subcontracting or joint ventures where another organization can be used to supplement the firm's own resources.

Project termination

With the benefit of hindsight it is always easy to conclude that an unsuccessful project should have been terminated earlier. It might be assumed that a well-designed system for continuous monitoring combined with a good evaluation system will give early warning of changes in the technical or commercial viability of a project. But it must be recognized that much of the information on which a decision must be based remains uncertain until late into a project. Balachandra [9] suggests fourteen questions which must be addressed.

Whatever the system adopted there are a number of factors which cloud the issue:

1. A sunk cost mentality (see Chapter 5) — stick with the existing project because of the investment already made.
2. New project euphoria — abandon the old in favour of the new which will always look more attractive on paper.
3. The will to win — all projects run into some sort of trouble; it is tempting to think that they can be overcome by determination and perhaps a little luck.
4. Absence of forecasting — the project is inherently sound and there is a well-defined market need, but unfortunately the timing is wrong.

In the final analysis the decision must be based on good judgment in the context of:

1. Avoiding the selection of projects which appear only marginally attractive, for the almost inevitable erosion of their desirability as time progresses will lead to their eventual termination.
2. A sound monitoring system to ensure the best possible information on which to base the decision whether or not to terminate.
3. Commitment at all levels including top management (for large projects), marketing and a project champion who can tip the scales in favour of ultimate success for a project which might otherwise fail through lack of purposeful management.

Summary

It is essential that no R & D funds are allocated to projects until there is reasonable confidence that a successful technological solution can be translated into a product that can be marketed profitably. Market considerations must dominate every R & D selection decision. If market criteria cannot be satisfied, the project is doomed from the outset.

Having established that a market opportunity appears to exist a wider examination of a range of interlocking internal and external factors is necessary in order to ensure that no significant barriers exist. When satisfied on this count the final stage of evaluation can then be undertaken. This must be a systematic evaluation of the project in relation to a comprehensive and detailed set of criteria.

A wide range of factors was examined which have a bearing on the ultimate success of an innovation. Many of these were unquantifiable and highlighted portfolio considerations in R & D, production, marketing and financial investment. They underlined the need for careful comprehensive evaluation throughout the life of a project undertaken by non-technological as well as R & D managers.

This analysis ensures that:

1. Every factor having a bearing on the total business impact of the project is examined explicitly.
2. Projects are rejected which fail positively to meet one or more essential criteria.
3. The need for additional information is revealed.
4. There is a common basis for comparison between projects.
5. A data reference point exists so that the effect of additional information on the viability of the project can be assessed accurately and quickly.
6. There is a procedure providing a basis for cooperation between R & D and other departmental managers.
7. The identification of apparent weaknesses leads to a constructive investigation of means for alleviating them.

Procedures for the overall evaluation of a project were discussed. It was concluded that a comprehensive checklist was essential. Nevertheless, the quality of managerial judgment cannot be replaced by any mechanistic technique. Moreover, the value of the evaluation can be vitiated in the absence of constructive managerial attitudes, thus the process is as important, if not more so, than the procedure itself.

References

1. Ruffles, P. C. 'Reducing the Cost of Aero Engine Research and Development', *Aerospace,* Nov. 1986.
2. Brunsson, N. 'The functions of project evaluation', *R & D Management,* **10**, No. 2, Feb. 1980.
3. Bright, J. R. 'Some Management Lessons From Technological Innovation Research', *Proceedings of National Conference on Management of Technological Innovation,* University of Bradford, 1968.
4. Kotler, P. *Marketing Management: Analysis, Planning and Control,* 2nd edn, Prentice-Hall, 1972 (Chapter 15).
5. Bayliss, J. S. *Marketing for Engineers* (Chapter 11), Peter Peregrinus, 1989.

6. Reekie, W. D. *et al. On the Shelf,* Centre for Industrial Innovation, 1971.
7. Becker, R. H. 'Project selection checklists for Research, Product Development, Process Development', *Research Management,* **XXIII**, No. 5, Sep. 1980.
8. Souder, W. E. and Mandakovic, T. 'R & D Project Selection Models', *Research Management,* Jul.–Aug. 1986.
9. Balachandra, R. 'Critical Signals for Making Go/No Go Decisions in New Product Development', *Journal of Product Innovation Management,* **1**, No. 2, April 1984.

Additional references

Augood, D. R. 'A New Approach to R & D Evaluation', *IEEE Transactions on Engineering Management,* EM-25, Feb. 1975.

Baker, K. G. and Albaum, G. S. 'Modelling New Product Screening Decisions', *Journal of Product Innovation Management,* **3**, No. 1, March 1986.

Baker, N. 'R & D Project Selection Models: An Assessment', *IEEE Transactions on Engineering Management,* EM-21, Nov. 1974.

Callon, M. *et al.* 'Tools for the Evaluation of Technological Programmes', *Technology Analysis and Strategic Management,* **3**, No. 1, 1991.

Krawiec, F. 'Evaluating and Selecting Research Projects by Scoring', *Research Management,* Mar.–Apr. 1984.

Ramsey, J. F. *Research and Development: Project Selection Criteria,* UMI Research Press, 1978.

Ryan, C. G. *The Marketing of Technology,* Peter Peregrinus, 1984.

Souder, W. E. 'A System for Using Project Evaluation Methods', *Research Management,* Sep. 1978.

5

Financial evaluation of research and development projects

One of the main advantages of trying to use formal selection and estimating techniques in research and development is not that significant, precise quantified results emerge, but rather that complex situations are shown to be capable of logical analysis, implicit assumptions about the various aspects are exposed, relevant historical evidence is marshalled, issues which require intuitive technical and commercial judgment are identified, and, in so far as quantification is possible, the dimensions of uncertainty are established on consistent statistical rules.

K. G. H. Binning

The need for cost effectiveness analysis

Investment in technology is only one of the many ways in which a business can employ its financial resources. Thus expenditure on R & D ought to satisfy the same financial criteria the organization applies to other uses of funds which have a long-term pay off. In this respect it is very similar to a capital investment. Having said this, however, the difficulty of preparing meaningful financial evaluations in a practical situation must be faced. A new capital investment can be analysed in great detail and reasonably accurate estimates for both the cost and the benefit are expected before an allocation of resources is authorized. An investment in R & D, however, usually yields a return only in the longer term. The size of this return is difficult to forecast. Nor is it easy to establish any quantified measure for the overall effectiveness of R & D investment in the past which could be used as an indication of what might be expected in the future. How can the profit contribution from a product developed several years ago be allocated between the R & D and marketing departments? The argument that without the R & D investment there would have been no new product, and consequently no profit from it, is some justification. But it is impossible to determine whether the same money could have been invested more profitably elsewhere or whether it should have been employed in the development of an entirely different new product. This is a topic that has received a great deal of attention in recent years [1] but it is doubtful whether it will ever be possible to derive any meaningful measures of R & D performance isolated from other organizational variables.

Even if it were possible to evaluate precisely the historical performance of an R & D department, it would be little guide to the present when the technology, the management and the control systems may be very different. It is the future which is important, not the past. Fortunately for the research director, he is rarely required to justify former allocations of funds. What concerns him is the level of his budget for future years. Nevertheless, the knowledge that it is impossible to measure his overall financial performance with any degree of accuracy should not lead him to lose sight of his primary objective, namely the provision of the potential for the organization's future profitability.

In retrospect, the R & D contribution will be seen to have been made by a few successful projects. The many failures will be largely forgotten. In the present, however, it is not known which projects will eventually prove the winners. The research director is concerned with a portfolio of projects from which he must aim to extract the maximum benefit. This portfolio is maintained by the addition of new projects to replace others which have been terminated or concluded successfully. In selecting these new projects their impact on the total portfolio must be considered as well as their individual merits. It might, for example, be advisable to reject a high-risk project, in spite of a high expected payoff from it, if the portfolio is small or if it already contains a number of other high-risk projects. In different circumstances that same project might be highly desirable.

The portfolio is continuously evolving. Its content at any time is largely determined by past decisions, but the correct balance within it must be maintained at all times. This is achieved by the selection of new projects. We have seen already in Chapter 4 that this decision should be influenced by many factors which cannot be expressed in quantifiable terms. Nevertheless, a financial evaluation is invariably one of the most important inputs to the selection process. Indeed many R & D directors rely almost entirely upon a financial analysis for their project decisions. In this they have to assess the project in relation to:

1. Some criteria of financial performance.
2. Other projects which meet the same criteria.
3. The impact the new project will have upon the balance within the portfolio.

In recent years a great deal of attention has been devoted to the development of mathematical techniques for project selection and analysing cost effectiveness. The value of these techniques is, however, dependent upon the quality of the data to which they are applied. This data consists of estimates for project costs and benefits, which can be modified by judgments of the likelihood of success expressed as subjective probabilities. Thus, before discussing the techniques, it is useful to examine the available evidence on the accuracy that can be expected in the information upon which these important decisions must be based.

Research and development financial estimates

Estimating accuracy

Substantial cost escalations and programme slippages associated with advanced technological projects have received considerable publicity. The serious financial consequences of some of these programmes have contributed to a reluctance by industry to invest further funds in high-risk technological innovations. A well publicized, but not untypical, example is the Concorde supersonic airliner for which the estimated R & D costs escalated almost linearly from £150 million in November 1962 to £1,065 million in June 1973. Similar escalations have been reported more recently in advanced defence projects in both the USA and Europe.

But these experiences are by no means exceptional. Does this indeed mean that the uncertainties and risks are such that companies should avoid projects involving advanced technology whenever there is a feasible alternative? If this thesis were accepted and acted upon the rate of technological progress would eventually grind to a halt. But the business concern investing in technology cannot afford to take risks which might ultimately destroy its financial viability, however desirable the results may be for society in the long term. It must assess the uncertainties realistically and only then decide whether the business risks are acceptable.

In evaluating the desirability of a new project the evidence from a number of studies into estimating accuracy must be borne in mind. These show clearly that:

1. Estimates are almost invariably optimistic.
2. The greater the technological advance involved, the larger the extent of underestimation.

These findings are not surprising. We shall now review some of these studies to see whether they can help us to assess the likely magnitude of estimating errors to be expected in a given industrial situation.

Although the problems of major programmes have been well publicized similar results have been found with smaller and less innovative projects. Norris [2] studied 475 British projects in four organizations. In this large sample of small projects it was found the project durations, 1.39 to 3.04 times the estimates, exhibited greater errors than the cost estimates, 0.97 to 1.51. Similar results were found by Mansfield [3] in a study of estimating accuracy in two American drug companies. In the first, an ethical drug manufacturer, the average factors were 1.78 for cost escalation and 1.61 for development period, whereas in the second, a proprietary drug firm, they were 2.11 for cost and 2.95 for duration. Mansfield noted that the greatest inaccuracies occurred with the more ambitious projects.

These studies were conducted in the 1960s and 1970s since when there has been little research reported in the literature. Many readers might argue that these problems are less serious today as R & D management has become more professional. There is little evidence to support this view which is not borne out by the author's own experience in several countries during the late 1980s (Table 5.1). Cordero (see note [1] earlier) comes to the same conclusion whilst stressing that the evaluation of innovation performance can yield valuable information for decision-makers.

Table 5.1 *Examples of estimating inaccuracies (1983–90)*

Project	Time factor	Cost factor	
A	1.9	1.8	
B	4.2	2.1	Large complex project
C	1.2	4.1	Major chemical project
D	1.3	3.1	
E	2.2	3.9	

It is interesting to note that in large projects, cost escalation appears more significant than slippage in the development time, unlike smaller projects where substantial delays seem more commonplace. It must be recognized, however, that these two factors are not independent variables. The underlying uncertainty of R & D projects relates to the amount of work which has to be undertaken to complete the project. Cost and duration are, therefore, to an extent interchangeable. An important programme which is slipping can usually be accelerated by 'crash' action, but at a cost. Beyond a certain time, delay is also likely to incur financial penalties for a variety of reasons − the tempo of work may decline; some of the overhead cost may be borne for a longer period; and changes in technology and market conditions may necessitate modifications to the project. But this does not mean that the duration corresponding to minimum development costs is necessarily the optimum for the business, since additional cost may be warranted in order to gain a market advantage which yields a profit greater than the cost penalty (Fig. 5.1). Thus a degree of 'crash' action may be the appropriate policy to adopt from the outset, rather than, as is so often the case, a reaction to recoup delays in the development programme.

The tradeoff between cost and duration may explain the differences we observed between large and small projects. The large projects normally have long development periods and it is often vital for the purchaser that delivery delays are minimized. Moreover, these projects are likely to be kept under detailed scrutiny by top management who demand the use of refined control techniques (e.g. PERT) to maintain the project on schedule, but at a cost. On the other hand, the small project is much more likely to be subject to interference from other projects in the portfolio, all of which are scheduled within an overall financial budget. From the viewpoint of effective management it is important that the implications of either cost or duration over-runs are considered explicitly.

Although the cost side of the benefit:cost equation is difficult to estimate,

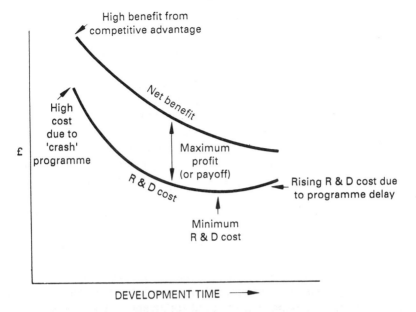

Notes: 1. Net benefit = Sales income – All costs attributable to the sales (i.e. production, marketing, overheads etc.) but excluding R & D
2. Profit (or payoff) = Net gain to the business from the project (net benefit – R & D costs)
3. Maximum payoff does not coincide with minimum cost R & D programme

5.1 Effect of project duration on cost and profitability

the benefit is liable to even greater errors. This is not surprising, for the sales forecasts must anticipate the buying response of the consumer several years ahead, in the absence of any specific price or performance information and not knowing what competitive products will be available. The great difficulties of forecasting the potential benefits are illustrated by the research of Beardsley and Mansfield [4] which indicates that large errors are found in the estimation of the profitability of a product until it has been on the market for up to five years after launch.

As a project progresses additional information should enable more accurate forecasts of the outcome to be made. In practice, although this does happen, it is often very late in the development programme before the estimates begin to approach the eventual figure.

Nor are we in a better position with the subjective probabilities for technical or commercial success. Mansfield's studies led him to conclude, '. . . the estimated probabilities of technical completion are of some use in predicting which projects will be completed and which will not. But they are not of much use, since even if they are used in such a way as to minimize the probability of an incorrect prediction, they lead to an incorrect prediction in about 30 per cent of the cases.'

These findings should not be surprising. Innovation is by definition the process of doing something which is new. Since it has not been done before a considerable element of ignorance is unavoidable, the difficulties that will be experienced cannot be known in advance and their cost implications estimated. It is almost an axiom that an accurate estimate implies that the project is not truly innovative. Although most technologists accept this as a generalization they still believe that their own estimates are not seriously flawed. It is interesting that much of the data in Table 5.1 was provided by researchers who examined the past performance of their own projects in an effort to refute this thesis and were surprised by their findings. In a study of seven major Japanese companies, the author found that these difficulties were accepted and none of them used financial estimates as a basis for their project selection decisions.

We may conclude that there is a considerable weight of evidence to indicate that the information available at project initiation is of extremely poor quality, so poor indeed that one may feel some justification in concluding that, if past experience is any guide, the data on which financial analyses are based is virtually useless for decision-making. Yet the need for financial analyses is central to R & D decision-making. One must accept that they will always be required by top management before important R & D investment decisions are made whether or not they are of any real value. Thus if the decisions are to be the best possible in a given set of circumstances, we must aim to ensure that:

1. All managers involved with project decision-making fully recognize the magnitude of the possible errors in the estimates, arising from the inevitable uncertainty of R & D.
2. Increased accuracy of estimation is achieved wherever possible.
3. Estimates are used in a way which is realistic given their possible errors.

Improving the accuracy of estimates

The magnitude of the estimating errors is not wholly attributable to the uncertainties inherent in the programme. Another reason may often be the lack of attention devoted to efforts to improve estimating performance. Perhaps this is a natural consequence of the knowledge that the unknowns preclude a high degree of accuracy, however much care is taken in preparing the estimates. Nevertheless, if important decisions are to be based on the estimates, it is worth exploring all possible methods for reducing the margin of error, and to spend money in doing so. A small expenditure at this stage may save a much larger sum later.

Estimating errors arise from five main causes:

1. Poor judgment by the technologist in estimating the cost and duration of specified tasks.

2. Changes to the project during development to overcome technical difficulties or alterations in the market.
3. Non-technical factors such as inflation.
4. New factors, e.g. environmental legislation.
5. Under-estimation of manufacturing and marketing costs.

The relative importance of these factors will vary with the characteristics of the project and the industry. In a short duration project, for example, the inflationary effect is likely to be relatively slight. It is in any case irrelevant to the evaluation of technological performance since inflation is changing the unit of measurement rather than the quantity being measured. Notwithstanding this, it is rare for the effects of inflation to be disaggregated from other factors when comparing a project outcome in relation to the estimates made at the time of approval. It should also be noted that programme slippage in an inflationary climate will increase the project costs in current money terms even when the actual resources employed are unchanged.

Design changes occasioned by altered market conditions are apt to assume appreciable proportions for the longer duration projects. Sometimes the market alters so significantly during development that a costly redefinition of the project becomes necessary. These changes are almost impossible to forecast. Nevertheless, in exercising a managerial judgment, one knows that programme modifications invariably involve additional expense and the longer the planned project duration the higher this cost is likely to be.

Thus it would appear that the most promising opportunities for improving estimating accuracy lie in developing the judgment of the estimator in respect of forecasting for specified tasks known to be necessary for the completion of the project.

A great deal can, however, be learnt from the past by studying historical data from earlier projects. If one of the parameters of interest, say development cost, is plotted for a number of projects as in Fig. 5.2, a distribution about a mean will be found. From this, the expected value for the cost of new developments may be inferred to average about 1.3 times the estimate, in the case of Laboratory A (Fig. 5.2). Because of the distribution of possible outcomes, this type of analysis is not of much use when applied to estimates for individual projects, but it does give a good indication of the likely performance of the whole portfolio, provided nothing has changed which could alter the accuracy of estimating. In this context it must be recognized that the establishment of a system for evaluating past performance may by directing attention to estimating, improve the accuracy of future estimates. For one of the major causes of poor performance is often the absence of any evaluation after the project is completed. No attempt is made to identify the causes of errors and, consequently, to learn from past mistakes. The absence of evaluation and awards or penalties for estimating accuracy may combine to create attitudes where little importance is attached to it. The aim must be to move the peak of the distribution (X in Fig. 5.2) towards the origin and to reduce the base of the distribution curve.

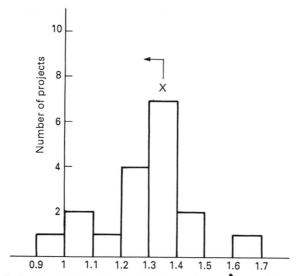

Ratio of actual to estimated R & D cost for new drugs (Laboratory A)

5.2 Analysis of historical estimating performance

Mansfield (see note [3] earlier) noted during his studies in the pharmaceutical industry that, 'As would be expected, the accuracy of a forecast seems to be related to the amount of resources devoted to its preparation.'

There are other ways in which lessons can be learnt from the past. No project is different in every respect from its predecessors. Although there are some areas where no previous experience exists to provide a guideline, this is by no means universally true, for much of the work is necessarily of a routine nature comparable with tasks performed on earlier projects. Examination of historical data should enable the costs to be allocated to various parts of the development programme (e.g. the test-bed running of a new aero-engine, the fatigue testing of airframe components, or environmental testing of new pigments). Frequently, analysis reveals an empirical relationship between some parameters of the product and the cost of a phase of the development programme (Fig. 5.3), which enables an estimate to be made for a new development by extrapolation subject to the proviso that the relationship may no longer be valid if the technology has changed substantially.

While analytical methods can assist where a task is reasonably well defined, they are of no help when there is no previous experience to draw upon. Here we are thrown back upon the manager's subjective judgment, and estimates will vary according to the skill of the individual, his attitudes, and his tendency to optimism or pessimism.

Epton [5] has suggested that individuals have a personality characteristic he calls the 'time horizon' within which their minds normally operate. Task completion dates are unconsciously predicted with the time horizon of the estimator, leading to underestimation of the duration of projects extending

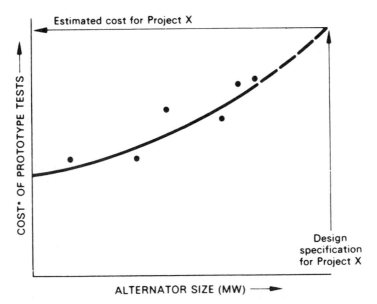

Estimated cost for Project X

COST* OF PROTOTYPE TESTS →

ALTERNATOR SIZE (MW) ──►

Design
specification
for Project X

* The cost must be expressed in a form unaffected by inflation, e.g. man-hours

5.3 Use of historical data in estimating future development costs

beyond this time horizon. This theory is, as yet, unproven and has been challenged, but Epton suggests that if further research validates it, we may, in his words, derive a useful management aid − 'If time horizons could be assigned to individuals, management might use them to allow for bias in time-estimates with some gain of precision of resource provision and budgetary control.'

Since judgment is involved in all estimating, the figures produced are bound to be a reflection of personal attitudes. Two people making estimates for the same item will make forecasts which are significantly different. To some extent this reflects their own tendencies towards optimism or pessimism. It is clearly undesirable to compare estimates for Project X by A, who is optimistic, with those for Project Y by B, who is pessimistic, when choosing between the two alternative projects.* Although it is impossible to eliminate these biases, their effect can be minimized by asking for estimates from several different persons who have sufficient knowledge of a project.

Psychological pressures also tend to lead to underestimation. The technologist preparing the estimates usually identifies closely with the project. He wants it to be selected. Thus, either consciously or subconsciously, he will produce figures

* Williams [6] carried out a study at the British Aircraft Corporation into the decision criteria used in evaluating R & D projects. He noted that, '. . . the scales derived from subjective estimates of probability of success while showing approximately the same number of levels, showed a marked difference in the specification of those levels. Thus some scales are linear while others were biased towards the higher values of probability.'

which present the project in the most favourable light. For he knows that the greatest hurdle to be cleared is initial selection; once money has been spent it becomes increasingly difficult to terminate. This may even cause deliberate distortions of estimates in order to mislead or, as Blake [7] of Stanford Research Institute writes: '. . . the kind of company management that attempts to tie profitability to particular R & D projects is almost certainly, in my opinion, going to force some measure of chicanery on the people who are responsible for putting the numbers together.'

But what happens to these estimates when presented? Managements do not learn from their past experience by adding a factor to make some allowances for underestimation. All the organizational pressures are in the reverse direction. The project is desired but the cost estimate is uncomfortably high. So, they conclude that the cost of the project must be reduced. This leads to re-examination of the estimates resulting in apparent economies which bear little relation to real cost reduction. This is often achieved by persuading the person who prepared the original estimates, usually the technical manager who knows the detail of the project best, to modify his previous judgment. Thus, in successive stages, the estimates are reduced to a more 'acceptable' figure by:

1. Removing from the programme elements which are 'desirable' but cannot be justified as 'essential'. In most cases these will be reinstated as development proceeds.
2. Reduction of contingency allowances. These cannot be justified since they are not specified in detail, being some allowance to cover uncertainties and unknown costs.
3. Reducing the estimates for the constituents of the programme. Since these represent the judgment of the manager he can usually be persuaded reluctantly to make some reductions.

These savings are mainly notional, resulting from wishful thinking. They will disappear as the project proceeds. Estimates must, of course, be scrutinized, but great care should be taken to ensure that the only outcome of organizational pressure is not distorted judgment. Self-deception leads to poor decisions.

Implications of estimating errors for project evaluation

The R & D manager is faced with a dilemma. He knows that financial criteria are an important element in project justification and that top management expect cost:benefit analyses to be made. On the other hand, he is aware of the poor quality of the information upon which the decision must be based. He cannot avoid the problem, but there are some actions which he can take to improve the accuracy of estimates:

1. *Take estimating seriously.* This means devoting resources to it.
2. *Study past performance.* These studies will indicate the order of accuracy

to be expected depending upon the technology, the type of project, and the degree of novelty.

3. *Analyse the causes of error.* This will improve performance by:
   (a) revealing relationships between cost and technological performance;
   (b) developing the judgment of the estimators who can learn from their past mistakes and who will also be more conscientious in estimating when they know their performance is being monitored.
4. *Avoid distorting judgment* by applying undue pressure to reduce estimates.
5. Wherever possible ask for *several independent forecasts* to minimize the effects of personal bias.

Whatever the achievable improvements in estimating performance, forecasts are still liable to the large errors inevitable when dealing with the unknown. Since they are the only data available they must be used, but used realistically. This means that:

1. Little reliance can be placed upon the absolute value of figures emerging from a financial analysis at the initial stages of a project. Nevertheless, they are a useful guide to the likely value of a project *relative* to others being considered at the same time.
2. Selections based upon them must be regarded as *limited* decisions. In this respect the R & D manager is in an advantageous position compared with other managers who have to carry out capital investment evaluations, for he is making only a limited initial commitment of resources. As the project proceeds he must ensure that continued effort is devoted to improving the quality of the financial estimates so that successive decisions are based upon data of increasing accuracy.

Risk as a factor in financial analysis

Every business decision involves risk. The greater the unknowns, the greater that risk will be. For a technological innovation there is always the possibility that the project will not prove feasible technically (or more commonly, the cost of solving technical problems becomes unacceptable), or a technically successful project fails in the market. The subjective evaluations of these risks as probabilities of technical (P_t) or commercial (P_c) success has been referred to earlier. The risk that the investment will be lost is $(1 - P_t \times P_c)$. However, this risk can be considered in two parts: firstly the R & D cost to feasibility $\times (1 - P_t)$ and later the cost after feasibility is established $\times (1 - P_c)$.

Where the investment is small, projects can be compared merely by modifying the expected benefit−cost ratio to take account of their overall risks, to give an index of merit:

$$\left(I = \frac{B}{C} \times P_t \times P_c \right)$$

although confidence in the value of the index is limited by the poor quality of the estimates for the individual factors. The business risk is the possible loss of its total investment in the project. This may be acceptable in view of the benefits expected from the project if it proves successful.

The considerations for a large project, absorbing a substantial amount of the company's resources, are different. The major business risk in this case may arise from the escalation of development cost to a figure beyond what the company's financial resources can support. This is a common cause of technology-induced business failure. It can happen even when the expected benefit:cost remains highly favourable in spite of the development cost increases. But it is not only an escalation of development costs which can cause liquidity difficulties. All causes of negative cash flow must be considered. With a successful project the greatest strain on liquidity may come from the need for working capital to support early sales.

The technique of risk analysis (Hertz [8]) enables quantitative analysis of the probabilities of possible outcomes in capital investment. A typical risk analysis curve is shown in Fig. 5.4; it can be seen that in this case there is a

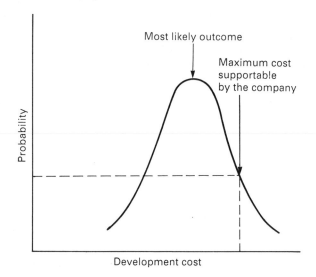

5.4 *Risk analysis for development cost*

significant probability of the development cost exceeding a level supportable by the company's financial resources, even though the expected value is acceptable. The subjective probability distributions for each of the factors relevant to a benefit:cost analysis can be combined to give an overall distribution for the expected cost:benefit.

The results of risk analyses can be used to compare the desirability of projects, not only in relation to each other but also in relation to the portfolio as a whole (Fig. 5.5). From this we can see that:

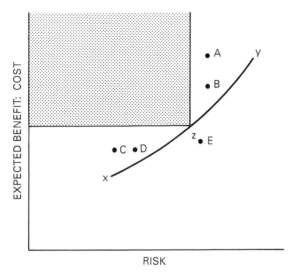

5.5 Evaluating risk in project comparisons

1. Project A is preferable to Project B, since A has the higher expected benefit:cost for the same associated risk.
2. Project C is preferable to Project D, since C has a lower risk for the same expected benefit:cost.

The line XY represents the minimum conditions considered essential for any project to satisfy. Thus Project E would be rejected on this criterion since the expected benefit:cost is too low for the risk involved. Consideration of the total portfolio leads to the establishment of criteria for its performance in relation to the needs of the business. This is represented by the shaded portion of Fig. 5.5 within which area any outcome is satisfactory, point Z being the minimum acceptable benefit:cost ratio for the portfolio as a whole. In these circumstances, Projects A and C would be selected in order to keep the risk within acceptable limits rather than A and B which would yield the highest expected benefit:cost for the portfolio.

The validity of the risk analyses is always in doubt because of the dubious nature of the data upon which they are based. Nevertheless, they can be of value to the research manager, provided he regards Fig. 5.5 as a rough pictorial representation rather than a precise assessment. This enables him to:

1. Consider the expected benefit:cost *v* risk for each individual project.
2. Represent graphically the benefit:cost *v* risk criteria which should be applied to projects.
3. Compare projects.
4. Establish overall criteria for the portfolio as a whole.
5. Select projects in relation to the needs of the portfolio.

While this is insufficient evidence for positive project *selection* it helps to categorize projects into three classes:

1. Those which should be *rejected* because they fall short of the criteria.
2. Those which can be *selected provided* they meet the non-financial criteria, because they exceed substantially the minimum criteria.
3. Those *marginally acceptable* subject to other considerations. In these cases it is usually wise to adopt the maxim, 'If in doubt, reject', since they are likely to become increasingly marginal as development proceeds.

Project selection formulae

Numerous formulae have been devised with the object of combining a number of the project selection criteria in one 'index of merit', thus enabling projects to be compared on a common basis. Ansoff [9] proposes two indices — a figure of merit (profit) and a figure of merit (risk) — in the derivation of which he combines all those factors he considers relevant to a project decision:

$$\text{Figure of merit (profit)} = \frac{(M_t + M_b) \times E \times P_s \times P_p}{C_d + J} \times S$$

where M_t = technological merit
   M_b = business merit
   E = estimate of total earnings over lifetime
   P_s = probability of success of project
   P_p = probability of successful market penetration
   S = strategic fit of proposed project with other projects, products and markets
   C_d = total cost of development including working capital and facilities
   J = a savings factor resulting from shared use of existing facilities and capabilities

$$\text{Figure of merit (risk)} = \frac{C_{ar}}{FM_p}$$

where C_{ar} = total cost of applied research
   F = total cost of extra facilities, staff, etc.
   M_p = figure of merit (profit)

In these formulae Ansoff combines both those factors for which a purely quantitative estimate can be made and subjective assessments for qualitative considerations such as strategic fit. While the formulae may give a satisfactory symbolic representation it is doubtful whether they provide a mechanism of much operational value. The judgments involved are so complex there is great danger of the formulae being used to apply a veneer of pseudo-quantification to support decisions which have already been taken on different considerations.

Most writers do not attempt to incorporate qualitative factors in their formulae. Hart [10], for example, derives an index based only on financial considerations and probabilities of success:

$$\text{Project index (benefit:cost)} = \frac{S \times P \times p \times t}{100 \ c}$$

where S = peak sales volume (£p.a.)
P = net profit on sales (%)
p = probability of R & D success (on a scale 0–impossible to 1–certain to succeed)
t = a discounting/timing factor (years)
c = future cost of R & D (£)

Similar formulae have been derived by the Industrial Research Institute [11], Carl Pacifico [12], Whaley and Williams [13], and many others.

The usefulness of these formulae depends upon the validity of the data to which they are applied. If the data is poor they are little better than descriptive representations of the problem. When applied to estimates, of the order of inaccuracy discussed earlier, they can do a positive disservice by concealing in a simple index the magnitude of the uncertainties. However, there are some types of R & D work where it is possible to assess both the benefits and costs to a high degree of accuracy. These are usually development projects. Sometimes where a project is aimed to improve the reliability of a production process or to reduce manufacturing costs it is possible to calculate the potential benefits within narrow limits, often more accurately than the estimates for the R & D cost. In such circumstances, the quantitative techniques are not only appropriate but can provide the main basis for project selection.

Thus in choosing the method of financial analysis the R & D manager must be careful to select what is appropriate to his own situation, bearing in mind that it must be consistent with the information available at the time.

The use of financial analysis

The apparent similarity between the financial analysis required in selecting an R & D project and the problem of appraising a major capital investment might suggest that the same evaluation procedures could be used in both cases. There are, however, important differences between the two. The financial data available when deciding upon a capital investment, for example, building a new factory, is considerably more reliable than for most R & D projects, particularly early in their lives. On the other hand, the R & D project has the advantage that, once initiated, it can usually be curtailed without severe financial penalties, whereas the capital project usually involves the commitment of a large proportion of the total investment at one decision which is only likely to be reversed if major unforeseen difficulties are encountered. The limited nature of the R & D project selection decision must be stressed. The inadequacy of the information is such that it only permits the weeding out of proposals which have little merit. Judgment must be reserved on those which are selected. The most that can be said in their favour is that no overriding reason for rejection has been identified, so they are allowed to proceed for the time being, but their future is not guaranteed.

During development a number of 'milestone' events will occur – the decision to build a prototype or pilot plant, or to construct a production facility. Each of these successive milestones is accompanied by the commitment of ever-increasing financial resources. Before proceeding from one phase to the next a re-evaluation is essential. As the size of the investment increases, so does the need for financial analysis. Thus one of the major objectives of an R & D programme must be the refinement of the information available when the milestone decisions are taken. The reduction of technical uncertainty, although important, is only one of the considerations. The aim must be to reduce the business uncertainty surrounding the future of the project.

It is necessary, therefore, to identify the major areas of uncertainty and gear the R & D programme to their resolution. Some areas may be critical for the profitability of the project, others relatively unimportant. Sensitivity analysis on the financial information will indicate how significant any departures from the estimates would be for the business success of the project. A few examples will show how the results of such an analysis can affect programme decisions. For example, we have already seen that in some circumstances the development cost itself may be the critical factor for a large project. In such a case it may be essential to concentrate effort on obtaining more precise information leading to improved estimates for the development cost.

One of the major problems affecting many advanced technological projects is the high level of R & D cost to be amortized over a small production volume. In such cases it may be desirable to reduce the development cost to a minimum, even when this is accompanied by a lower performance; provided it attains an acceptable level; the 'best' can be the enemy of the 'good'. This minimization of R & D cost cannot usually be achieved however, without paying the penalty of higher unit production costs, the effect of which is illustrated in the breakeven chart shown in Fig. 5.6. Comparison of the two projects in Fig. 5.6 shows that the minimum development cost alternative offers a lower breakeven level of sales and higher profit margins except at high sales volume. For computer manufacture, Arthur D. Little has estimated that doubling the sales volume halves the R & D cost/unit and reduces unit costs by 10–15%. On this basis they calculate that the minimum viable size for a computer manufacturer (assuming the lowest acceptable pretax profit to be 5%) is one-eighth that of IBM (pretax profits 26%) representing 7.8% of the world market. In this highly geared situation the profitability of a manufacturer competitive with IBM is very sensitive to development costs which should consequently be kept to the

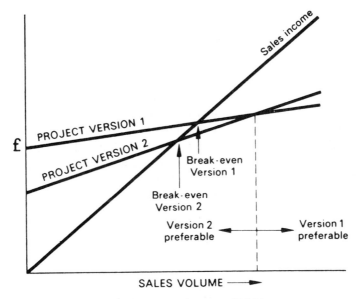

SALES VOLUME ⟶

Note: PROJECT VERSION 1 - high fixed cost (R & D)
　　　　　　　　　　　　　 low variable cost (manufacture)

　　　PROJECT VERSION 2 - low fixed cost (R & D)
　　　　　　　　　　　　　 high variable cost (manufacture)

5.6 Sensitivity of profitability to development costs for high technology/low sales volume projects

minimum acceptable, in spite of some consequent penalties in manufacturing cost.

The considerations may be entirely different for another project which leads to a product with a limited useful life. Several years after introduction a new technology, still in an early stage of development, is expected to result in an innovation which will replace it. This would suggest the desirability of devoting some resources to technological forecasting in order to estimate more accurately the date of obsolescence. But financial success may depend much more upon extending the useful life by aiming to introduce it on the market at the earliest possible date. Several alternative crash programmes might be evaluated leading to the selection of one where the additional development cost (dC) yields an increased benefit from early introduction (dB) such that $(B + dB):(C + dC)$ is a maximum and greater than $B:C$.

A further example is provided by a project aimed at a market currently satisfied by the established product of a competitor. Success will depend upon producing a product which can be sold more cheaply than the existing product or with a superior performance, but it is not certain which is considered more important by the consumer. Investment in further market research is needed before the objective of the project can be defined precisely. This would indicate whether the R & D programme should be concentrated upon performance, lower manufacturing costs, or some combination of the two. The correct decision may be critical for the success of the new product, since increased performance usually incurs a penalty in higher manufacturing cost.

The approach discussed above implies flexibility. Although each allocation of resources relates to a specific objective (i.e. the refinement of information), the exact specification for the project is open to change in the light of additional information, as is the decision whether the project should be allowed to continue. Much of the information needed is gained from the technical progress of the project, but it must not be forgotten that the technical R & D, important as it is, can only provide some of the answers required for decision-making.

Financial analysis can thus be seen to be a continuing theme throughout R & D. A realistic approach to the problem is based upon:

1. Acceptance that initial selection is a *limited decision*.
2. Identification of areas where business success is *sensitive to errors* in the estimates.
3. Allocation of resources to *obtaining information* in the sensitive areas.
4. Using the information obtained *to decide*:
    (a) whether the project should be abandoned;
    (b) whether the project should be redefined;
    (c) where additional expenditure should be allocated to improve the data critical for decision-making at the next milestone.

Financial analysis at a milestone event

Termination of a project is always difficult, but never more so than after it has reached an advanced stage, for example after prototype or pilot plant testing. By this time the estimates of total development costs will almost invariably have increased and its future earnings potential may also have decreased. The project will have developed a momentum of its own and a decision to cancel may be thought to be an admission of an error of judgment in initiating it. With the new information the selection decision may be regretted, but it does not necessarily imply that it was a bad decision given the data available at the time it was made. No blame should attach to the R & D manager making that decision if he had carried out a systematic evaluation. Nevertheless, the R & D manager may be reluctant to terminate because he feels that this might be interpreted as a failure on his part and also because he is likely to have identified himself with the success of the project. It is difficult to be objective when too close to the project.

We may consider the re-evaluation at a milestone event by reference to the cash flow curves (Fig. 5.7). At time *t* the profile of the cash flow curve (2) is less attractive than when the project was chosen (1). In deciding whether to proceed or terminate it is necessary to carry out a new calculation of the benefit:cost ratio expected from the project. It is important to remember that the figures relevant to this decision are the currently expected benefit and the

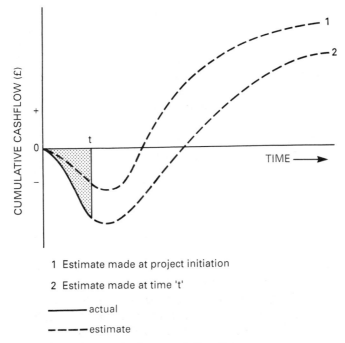

1 Estimate made at project initiation

2 Estimate made at time 't'

——— actual

———— estimate

5.7 Cost : benefit for project termination — cash flow diagrams

future expenditure required to completion. The past expenditure, represented by the shaded portion of the graph is what accountants term a 'sunk cost'. It is money which has been spent and is, for the most part, irrecoverable, and is thus irrelevant to a decision in the present. Many R & D managers have difficulty in accepting this fact, but it is as incorrect to terminate the project at time *t* on the basis of total benefit:cost as it is to continue because of the past expenditure.

We can illustrate the rationale of this by examining an example. The financial figures might be as follows:

	Cost to date	Estimated future cost to completion	Estimated benefit
Time 0	0	£10,000	£20,000
Time *t*	£8,000	£6,000	£12,000

Note:
The figures used in the evaluation must be factual cash numbers which are not necessarily the same as the accounting numbers.

It can be seen that the expected benefit:cost was 2 at time 0 and also 2 for the future expenditure required at time *t* (i.e. £12,000:£6,000). Thus the correct decision at time *t* should be to continue with the project even though it will result in an eventual loss of £2,000, for the £8,000 is a sunk cost. Furthermore, the existing project may well be preferable to an alternative new project with an identical expected benefit:cost since the expenditure of £8,000 should have resolved many of the uncertainties and the future risk is correspondingly reduced. In many cases, however, confidence in the cost estimates may be lower due to the technical problems encountered; this would reduce the subjective probability for technical success and lead to termination.

Thus financial analysis can be of greater value in a decision whether to terminate or continue than at project selection. Emotion plays a much larger role when a project has been worked on for some time, but fortunately the data for financial analysis is by then improved allowing a more objective basis for decision-making.

Allocation of resources between research and development

Examination of a typical cash flow curve indicates that the cost of applied research is generally low compared with the later stages of a project. The major costs are incurred in the construction and testing of prototypes and pilot plants and later in the facilities for manufacturing. In terms of time, however, each of the three stages may cover approximately the same duration. Where total R & D duration is to be kept to a minimum some overlap between the stages may be thought desirable, e.g. starting construction of a manufacturing facility before prototype testing is completed. Such a policy obviously increases the financial risks if, for example, severe development trouble is experienced during a late stage of the prototype test programme.

The risk is minimized when the time allowed for collecting the information needed for making the next stage decision is increased. But this delays the completion of the project usually reducing its potential benefit. Moreover, some of the uncertainties can only be resolved by information obtained during the later stages of development. In addition, some of the information may change during a long development programme. For example, the operational economics of the Concorde airliner were affected by factors which could not have been forecast when the programme commenced, such as price competition in airline fares, decreased purchasing power of the airlines brought about by an economic recession, and a sharply rising trend of aviation fuel prices.

Thus critical decisions cannot be unduly delayed while awaiting information. There is an inevitable price to be paid for certainty, and benefits can be sacrificed in the search for risk reduction. Consequently, at each milestone a decision must be made whether or not to proceed on the basis of imperfect information. This imposes a limit to the time available for applied research. The problem is to use this time as effectively as possible.

Much of the work carried out in applied research is sequential and cannot be accelerated. But this is not always the case and it is frequently possible to carry out additional supporting research or parallel programmes given the availability of resources (Fig. 5.8). The extra cost may be negligible compared with the benefits arising from the shortening of the overall R & D programme when the research item is on the critical path (i.e. savings in R & D duration which directly affect the overall project duration).

There is thus a strong argument for bringing to bear as much effort as it is possible to employ on applied research, first to improve the quality of information upon which subsequent heavy investment is based, and secondly to buy time at what may be a relatively low price.

Projects

A - Estimated cost, £10000; Duration, 6 months; Probability of success 0.6
B - Estimated cost, £9000; Duration, 9 months; Probability of success 0.8

Alternative programmes

1 Select 'A' - If 'A' fails, select 'B'
2 Select 'B' - If 'B' fails, select 'A'
3 Initiate 'A' and 'B' simultaneously

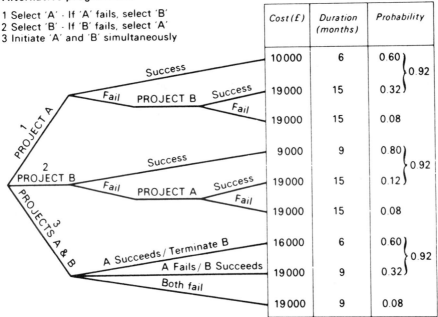

Cost (£)	Duration (months)	Probability	
10000	6	0.60	0.92
19000	15	0.32	
19000	15	0.08	
9000	9	0.80	0.92
19000	15	0.12	
19000	15	0.08	
16000	6	0.60	0.92
19000	9	0.32	
19000	9	0.08	

Note: Programme 3 ensures the outcome is known within nine months at the latest, but the cost cannot be less than £16000

5.8 Sequential or parallel approaches to the solution of an R & D problem

Financial discounting techniques (DCF)

In recent years the use of discounted cash flow techniques (DCF) in the evaluation of capital projects has become widespread. These are based upon the fact that cash flows have a time dimension. A pound in the hand now is worth more than a pound in a year's time because of the financial return it could have earned through investment during the intervening period. In times of high interest rates or where the opportunity cost for a company's investment is high, this factor is of significant importance. For example, where the rate of return is 10% the present value of £1 earned in ten years' time is only 38.6p,* reducing to only 9.2p in twenty-five years.

The effect of discounting on the financial analysis of two projects is illustrated in Table 5.2. Both involve the same investment, £80,000 spread over five years and returns spread over six years. Without discounting, Project B appears the more desirable since the total earnings over the product life of £140,000 are £10,000 greater than from Project A. However, when the figures are discounted at 10% the desirability of the two projects is reversed due to the different cash flow profiles. This arises because the project B costs are incurred earlier and

Table 5.2   *Effect of financial discounting on project selection*

Year	Discount factor	Cost	Discounted cost	Benefit	Discounted benefit
Project A (10% discount rate)		£	£	£	£
1	0.909	10,000	9,090		
2	0.826	20,000	16,520		
3	0.751	20,000	15,020		
4	0.683	20,000	13,660		
5	0.621	10,000	6,210	20,000	12,420
6	0.564			25,000	14,100
7	0.513			25,000	12,825
8	0.467			25,000	11,675
9	0.424			20,000	8,480
10	0.386			15,000	5,790
		80,000	60,500	130,000	65,290
Project B (10% discount rate)					
1	0.909	20,000	18,180		
2	0.826	20,000	16,520		
3	0.751	20,000	15,020		
4	0.683	10,000	6,830		
5	0.621	10,000	6,210	10,000	6,210
6	0.564			20,000	11,280
7	0.513			20,000	10,260
8	0.467			20,000	9,340
9	0.424			35,000	14,840
10	0.386			35,000	13,510
		£80,000	£62,760	£140,000	£65,440

	Benefit:cost undiscounted	Benefit:cost discounted
Project A	£13,000:£80,000 = 1.62	£65,290:£60,500 = 1.09
Project B	£140,000:£80,000 = 1.75	£65,440:£62,760 = 1.04

* Details of the alternative discounting techniques and discount tables are given in most texts on management accounting.

the benefits received later. The value of both projects is significantly reduced by discounting because the investment occurs in earlier years than the benefits. This effect would be amplified with a higher discount rate. An opportunity rate of 20% or more may be appropriate for a rapidly expanding company when its ability to invest in highly profitable additional areas is limited solely by the availability of funds. Furthermore, discounting increases the benefit from the timely completion of projects and consequently the penalties for project duration over-run.

A further problem arises with the application of DCF to projects for which the payoff lies in the distant future. Where the discount rate is high it becomes almost impossible to justify any expenditure with a payoff more than ten years hence (e.g. the present value of the £1 in ten years at a discount rate of 20% is only 16.2 pence). On this basis it would not be possible to maintain a continuous applied research programme in industries where the gestation periods are long. Yet if the company intends to remain in that business it must continue to invest in its technology. For this reason some research directors have rejected the use of discounting techniques. A report to the British Cabinet Office states, 'Falling profitability, uncertainty about the future and high hurdle rates of return combine to block off prospects for most projects. However, we think that consideration should be given to the appropriateness of conventional discounted cash flow (DCF) and payback calculations when the future role of the company is at stake.' [14] The problem disappears if one remembers that the business objectives of a company are:

1. Survival.
2. Earning a profit.
3. Other objectives such as social responsibility.

Survival is of first priority and R & D investment to ensure survival can be justified on those grounds alone, irrespective of the financial return. The investment of these relatively minor long-term funds can be likened to an insurance policy which must be paid as the price of remaining in business; this exception does not invalidate the relevance of DCF in all other financial analyses affecting R & D.

Application of operational research techniques

Management scientists have devoted a great deal of attention to the application of operational research techniques in the evaluation of R & D projects. In spite of the many research studies reported in the literature (see under additional references), there have been very few examples of their application or continued use by R & D management. The reasons for this can be seen from the studies of the accuracy of estimates and subjective probabilities. Where the data inputs are subject to such gross errors there is little likelihood of meaningful results emerging from the application of sophisticated mathematical processes.

There are, however, some situations where it is possible to estimate both the costs and benefits with a sufficiently high degree of accuracy to permit a relatively good approximation to the eventual benefit:cost to be made. In such circumstances it is often the cost rather than the benefit which is the more difficult to calculate. An example of this would be an R & D programme to double the mean time between failure of an aircraft component. The current reliability data will have been recorded and the cost associated with a failure (i.e. replacement cost of the component, labour cost in fitting, lost opportunity cost of aircraft availability, etc.) can be computed very accurately. It is also likely that the R & D cost to effect the improvement can be calculated fairly precisely.

The computed benefit:cost may seem attractive, but the project may appear less desirable when considered in relation to its effect upon the portfolio due to an excessive use of a scarce resource. Thus the problem is reduced to one of resource allocation. Bell and Read [15] illustrate this problem in a simple

Table 5.3 *Effect of resource constraints on portfolio selection*

(a) Projects incurring costs under one budget

Project No.	1	2	3	4
Cost	28	40	12	16
Benefit	35	60	30	28
Benefit:cost	1.25	1.5	2.5	1.75

Available budget   = 68
Selection of projects = 3,4,2
Total benefit   = 118

(b) Projects incurring costs under two separate budgets

Project No.	1	2	3	4
Wages cost	24	30	10	4
Capital cost	4	10	2	12
Benefit	35	60	30	28
Benefit:cost	1.25	1.5	2.5	1.75

Wages budget   = 54
Capital budget = 14
Total budget   = 68

Selection of projects	3 and 4	1 and 2
Total benefit	58	95
Benefit:cost	58:28 = 2.05	95:68 = 1.45

Source: Bell and Read, 'The Application of a Research Project Selection Method', *R & D Management,* **1**, No. 1, Oct. 1970.

example (Table 5.3). In this example, reference to the total budget indicates the selection of Projects 3, 4 and 2. However, when the budget is analysed into wages and capital costs it is seen that the maximum benefit is achieved by a combination of Projects 1 and 2. It is this type of resource allocation problem which linear programming is designed to tackle.

Bell and Read describe the development of a linear programme for a multi-project portfolio which has been used to aid project selection in the improvement of UK power-station performance. A number of constraints are built into the model in addition to the division between wages and capital costs; these include availability of different technical skills, transfer or recruitment of staff, alternative project versions and equipment constraints. They stress, however, that the linear programme is only an aid to managerial judgment. It lends itself to sensitivity analysis. For example, it is possible to evaluate the effect of budget changes on the choice of projects and the associated benefits. Due to organizational changes in the laboratory the use of this model was subsequently discontinued. It did, however, leave a longer term effect upon the staff employed in project selection; although no longer using the model they continued to consider the factors concerned more explicitly than they had done in the period prior to its introduction. It appears, therefore, that an appreciation of the linear programming technique may have a valuable educational effect even where it cannot be applied operationally. However, to return to the subject of forecasts, this technique can only be of real practical value when there is confidence that the estimates sensibly reflect the resources which will eventually be found necessary.

Many forms of mathematical model have been developed and reported in the academic literature. However, research studies in the USA (Liberatore and Titus [16]), the UK (Watts and Higgins [17]) and Europe (Fahri and Spatig [18]) indicate that they have had little or no application in industrial R & D decision-making. We may conclude that, whilst they can lead to a better understanding of the underlying relationships, the difficulty of obtaining sufficiently accurate estimating data invalidates them as a practical management tool. In the future, however, the development of expert systems may overcome some of these difficulties.

The main reason for this slow rate of adoption of quantitative techniques we have already seen to be the inadequate nature of so much of the data. Thus in many situations little is to be gained from detailed quantitative analysis. Unfortunately, there are grounds for believing that rejection of the technique in its detailed application is often accompanied by a rejection of the concepts behind the technique. Thus it may be realistic to conclude that sophisticated risk analyses are meaningless because of the quality of the subjective judgment, but it is certainly misguided to make no attempt to consider risk and uncertainty explicitly when making important project decisions.

In some cases, where quantitative techniques have been adopted and later abandoned, the reasons given have included the amount of scarce managerial time required and the inability of the system to adapt to changing organizational strategies. Nevertheless, the thought disciplines imposed upon the staff during the time the techniques were in use is claimed to have had a lasting effect.

Communication problems — R & D and finance

Ideally the R & D manager should possess sufficient knowledge to be able to communicate with financial staff in the language they understand. This presupposes some familiarity with capital appraisal, risk analysis, discounting techniques, sensitivity analysis and accounting procedures. On the other hand it is highly desirable that the accountant or financial manager has some understanding of technology and sensitivity to the process of innovation. This mutual understanding is unusual and conflict leading to poor decisions is often the consequence.

Some years ago the author was invited to visit the laboratory of an electronic multinational; after a two-hour tour in which the work was fully and enthusiastically described my host introduced himself as the chief accountant. This laboratory had an outstanding record and an absence of the recriminations between R & D and finance which one finds all too often. Unfortunately this is a rare example. What can be done to encourage such interaction? There are two initiatives open to the technologist:

1. *Learn his language.* This may not be easy; standard accounting texts can be hard reading and few address R & D problems. There are, however, a few books written especially for the technologist, e.g. Bisio and Gastwirt [19].
2. *Involve the finance staff.* Occasionally invite them to look around the laboratory and explain to them what you are doing in terms they can understand. You are likely to receive a pleasant surprise.

Summary

Examination of financial forecasts, made at the time of R & D project selection, reveals that they are almost invariably subject to gross errors when compared with the development costs which were subsequently incurred or the benefits eventually derived from them. Similar inaccuracies have been noted in estimates for project duration and the subjective probabilities for technical and commercial success. Such forecasts are consequently of limited value in project selection. Nevertheless, the effectiveness of R & D expenditure is so dependent upon the financial performance of projects that some form of financial analysis cannot be avoided. But no purpose is served by analyses which are unrealistic and give rise to misleading decisions.

Decision-making can and should be improved by increasing the accuracy of estimates. The scope for this is, however, limited by the uncertainties inherent in an activity primarily concerned with the unknown. Nevertheless, improvements can be made but this will only happen if the accuracy of estimating is made a topic for serious study within the firm.

The greatest contribution to decision-making comes from the recognition that project selection is a limited decision. Subsequent decisions will be based on improved information. For the most effective use of resources it is necessary to identify those factors to which the financial success of the project is sensitive, and to channel resources into reducing the uncertainties in those critical areas before committing major investment for future development. This may also result in adjustments to the objectives and timing of a project.

Risk is another important factor to consider explicitly. It was seen to enter the calculation in three ways — in relation to the uncertainties associated with the individual project; the risk of the portfolio as the whole; and the potential risk to the business from failure or cost over-runs.

Project selection formulae and quantitative techniques of evaluation were examined. It was seen that their limited use by R & D management was largely explained by the inadequate quality of the input data. There are, however, opportunities for the application of techniques such as linear programming, particularly in resource allocation and portfolio balancing, in those situations where it is possible to forecast both the cost and benefits with some degree of accuracy.

In the future it is to be hoped that greater attention will be paid to improving forecasting accuracy — by retrospective analysis of the causes of estimating errors, further study into the behavioural influences affecting the estimator, and the development of improved estimating techniques. Only then will there be a more satisfactory basis for a meaningful extension of the use of quantitative evaluation techniques leading to improved financial evaluation of projects and increased effectiveness of R & D investment.

References

1. Cordero, R. 'The Measurement of Innovation Performance in the Firm: An Overview', *Research Policy,* **19**, No. 2, Apr. 1990.
2. Norris, K. P. 'The Accuracy of Project Cost and Duration Estimates in Industrial R & D', *R & D Management,* **2**, No. 1, Oct. 1971.
3. Mansfield, E. *et al. Research and Innovation in the Modern Corporation,* Macmillan, 1972.
4. Beardsley, G. and Mansfield, E. 'A Note on the Accuracy of Industrial Forecasts of the Profitability on New Products and Processes', *Journal of Business,* **51**, No. 2, 1978.
5. Epton, S. R. 'The Underestimation of Project Duration. An Explanation in Terms of Time Horizon', *R & D Management,* **2**, No. 3, June 1972.
6. Williams, D. J. 'A Study of a Decision Model for R & D Project Selection', *Operational Research Quarterly,* **20**, Sept. 1969.
7. Blake, S. P. 'Some Hypotheses on the Management of Research and Development', in: *Management and Technology,* **1**, Mencher, A. (ed.), SPF Special Publications Series, 1972.
8. Hertz, D. and Thomas, H. *Risk Analysis and its Applications,* Wiley, 1983.

9. Ansoff, I.H. 'Evaluation of Applied Research in a Business Firm', in: *Research, Development and Technological Innovation,* Bright, J. R. (ed.), Irwin, 1964.
10. Harg, A. 'A Chart for Evaluating Product Research and Development', *Operational Research Quarterly,* **17,** 1966.
11. Heyel, C. *Handbook of Industrial Research Management,* 2nd edn, Reinhold, 1968.
12. Pacifico, C. *A.M.A. Research, Study 89,* Dean, B. V. (ed.), American Management Association, 1968.
13. Whaley, W. M. and Williams, R. A. 'A Profits-oriented Approach to Project Selection', *Research Management,* Sept. 1971.
14. *Industrial Innovation,* Cabinet Office: Advisory Council for Applied Research and Development, HMSO, December, 1978.
15. Bell, D. C. and Read, A. W. 'The Application of a Research Project Selection Method', *R & D Management,* **1,** No. 1, Oct. 1970.
16. Liberatore, M. J. and Titus, G. J. 'The Practice of Management Science in R & D Project Management', *Management Science,* **29,** 1983.
17. Watts, K. and Higgins, J. C. 'The Use of Advanced Management Techniques in R & D', *Omega,* **15,** 1987.
18. Fahri, P. and Spatig, M. 'An Application-oriented Guide to R & D Project Selection and Evaluation Methods', *R & D Management,* **20,** No. 2, Apr. 1990.
19. Bisio, A. and Gastwirt, L. *Turning Research and Development into Profits,* Amacom, 1979.

Additional references

Beattie, C. J. and Reader, R. D. *Quantitative Management in R & D,* Chapman & Hall, 1971.

Cooper, R. G. 'The Components of Risk in New Product Development', *R & D Management,* **11,** No. 2, Apr. 1981.

Foster, R. N. 'Estimating Research Payoff by Internal Rate of Return Method', *Research Management,* Nov. 1971.

Freeman, P. and Gear, A. E. 'A Probabilistic Objective Function for R & D Portfolio Selection', *Operational Research Quarterly,* **22,** No. 3, Sept. 1971.

Gee, R. E. 'A Survey of Current Project Selection Practices', *Research Management,* Sept. 1971.

Libik, G. 'The Economic Assessment of Research and Development', *Management Science,* **16,** No. 1, Sept. 1969.

Lockett, A. G. and Freeman, P. 'Probabilistic Networks and R & D Portfolio Selection', *Operational Research Quarterly,* **21,** No. 3, Sept. 1970.

Mansfield, E. *Industrial Research and Technological Innovation – An Econometric Analysis,* Longman, 1969.

Meyer-Krahmer, F. 'Recent Results in Measuring Innovation Output', *Research Policy,* **13**, No. 3, June 1984.

Ministry of Technology, *Report of the Steering Group on Development Cost Estimating,* HMSO, 1969.

Patterson, W. C. 'Evaluating R & D Performance at ALCOA Laboratories', *Research Management,* **XXVI**, No. 2, Mar.–Apr. 1983.

Souder, W. E., Maker, P. M. and Rubinstein, A. H. 'The Successful Experiments in Project Selection', *Research Management,* Sept. 1972.

Thomas, H. 'The Debiasing of Forecasting in Research and Development', *R & D Management,* **1**, No. 3, June 1971.

Thomas, H. 'Some Evidence on the Accuracy of Forecasts in R & D Projects', *R & D Management,* **1**, No. 2, Feb. 1971.

6

Research and development programme planning and control

The process of planning, which assumes the rational view, may be useful even though plans are bound to be inadequate. The formulation of objectives for technical effort provides direction for the effort and a stimulus for action, even though the objectives will have to be modified in the light of discoveries made in the process. There is a utility in the formulation of such objectives, provided this flexibility is allowed and expected, otherwise they strangle invention.

Donald Schon

The elements of the problem

R & D management works within continually changing conditions which necessitate a constant review of the R & D programme as a whole, as well as a periodic re-evaluation of its constituent parts in respect of the contribution each makes. This contrasts with the problems encountered by most other managers where it is usual for their decisions, once taken, to have a high degree of permanence. Although they will need to modify their plans from time to time in response to unanticipated events, they will normally be in a position to forecast the types of change which might be necessary and to prepare contingency plans. Moreover, the major uncertainties they have to contend with arise usually from changed circumstances outside the company's immediate control – new government regulations, changes in the rates of interest or taxation, a fall in demand, or an increase in raw material prices. But, for the R & D manager, both internal and external uncertainty surround his every action. At any time he may encounter an unexpected technical problem which may require him to delay or even abandon a project, or to reallocate resources not only to the project concerned but throughout the laboratory due to the interaction between projects drawing their resources from the same pool. Or, he may receive a new assessment of consumer needs or likely demand which might necessitate a re-evaluation of the project's viability.

Thus, when considering the planning and control of an R & D programme, we must always remember that we are concerned with the management of a dynamic situation. Before any decision or plan is made, the best information available at the time must be taken into account. However, fresh information is continually becoming available, any of which might invalidate a decision made only a short time previously. Therefore, any system of R & D planning and control must be sufficiently flexible to accommodate frequent modification without creating a total dislocation of the working programme. This does not mean that formal control systems cannot be installed but it does mean that the dynamics of the situation demand a higher degree of managerial attention than do other activities, many of which can often be left to operate unaltered for long periods without review.

In earlier chapters we have alluded to many of the considerations the R & D manager must assess in his decision-making. We must now consider how he can incorporate them in an operational system of planning and control for the day-to-day management of his programme. It must be assumed that these decisions will be taken within a framework of departmental strategy and objectives previously formulated.

This chapter is concerned with an examination of the problems the manager will face and how they can be accommodated within a planning and control system. The discussion is confined to the principles involved, rather than to a detailed examination of the application of specific techniques beyond the scope of this book and which can be found in standard textbooks on R & D administration. Attention will be focused on the following essential elements of this activity:

1. Project definition.
2. Portfolio planning.
3. Project planning.
4. Control information and control techniques.
5. Resource management.
6. Staff motivation.

Project definition

Every project must start with a clearly defined statement of the objective it is intended to achieve and against which its success can be measured. This is usually called the 'project definition'. Since ultimate success lies in the market-place, the objectives must be clearly defined in terms of market need, albeit modified by an assessment of these needs in terms of what is likely to be technically achievable in practice. Normally, however, there is a choice. Most markets are not monolithic and usually they can be broken down into segments, each of which satisfies the needs and tastes of a particular group of potential customers. The main characteristics of a market segment can be defined in terms of four interrelated variables – size of market, acceptable price, technical performance requirements, and time. Similarly, most products can be offered in a variety of forms differentiated by performance, cost and the date first available on the market. The possible relationships for a hypothetical product and market are illustrated in Fig. 6.1.

The example in Fig. 6.1 indicates three market segments and the characteristics of products to satisfy them. There are, of course, an infinite number of possible alternatives which could be considered. While this approach is adequate conceptually, it may be of little help in defining the target market at the time of project initiation since the form of these relationships will be imperfectly understood. Nevertheless, it is necessary to form an opinion of how the market is likely to segment, based as far as possible on market research information, in order to select that segment which offers the most promise from corporate and technical considerations and to define the parameters of the project so that they are internally consistent with the needs of the chosen segment.

Particular attention should be paid to evaluating what technical performance the market segment is likely to demand. For frequently, the enthusiasm of the technologist to achieve a high level of performance leads to overambitious technical targets not based upon a realistic assessment of what the customer actually requires. Excess performance almost inevitably carries a penalty in the cost of both R & D and manufacturing, and also in development time, thereby eroding the potential profitability of the product. It may seem an intelligent anticipation of future demands to aim for a performance in excess of what is currently specified by making some allowance for the needs of second generation or 'stretched' versions of the product. This should not be a unilateral decision of the technologist. It affects product planning and costs money. There may be sound reasons for making provision for higher performance needs at a later date. If so, it should be a conscious decision taken after due consideration by both marketing and technical departments.

When the project is first defined it is usually necessary to specify the product characteristics within limits imposed by the uncertainties. But with the improved information obtained as the project progresses, these limits can be reduced until they match closely the anticipated requirements of a selected group of customers whose needs should also be more accurately determined during the period of product development (Fig. 6.2). Thus the project definition becomes more precisely focused on a specific market need.

At the stage of initial project definition it is essential to concentrate attention upon the *problem* (i.e. the market needs in terms of human satisfaction) rather than on the *solution* (i.e. preconceived ideas of what the final product should be like). For this is the time when alternative solutions should be sought actively and creative thought harnessed to the problem. The sequence must be − *what* is to be achieved?, *how* can it be translated into a practical form?, and only then, *which* alternative holds the greatest promise? After an exhaustive search and the selection of the most attractive project concept, attention can then be turned to the technical details and the specification of the programme of work to be undertaken.

Thus the project definition needs to be concise and specific, but couched in

1 - Market relationships

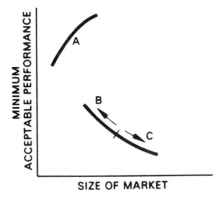

A - Small market for high
performance product

B - Medium market for medium
performance product

C - Mass market for low performance
product

2 - R & D relationships

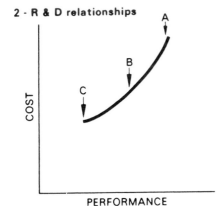

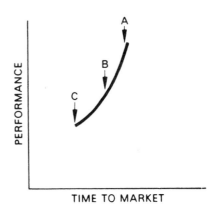

Note: These curves are illustrative and do not represent general relationships. Wide
variations will be found, depending upon the nature of the product, the technology
and the market.

6.1 Relationship between variables in project definition

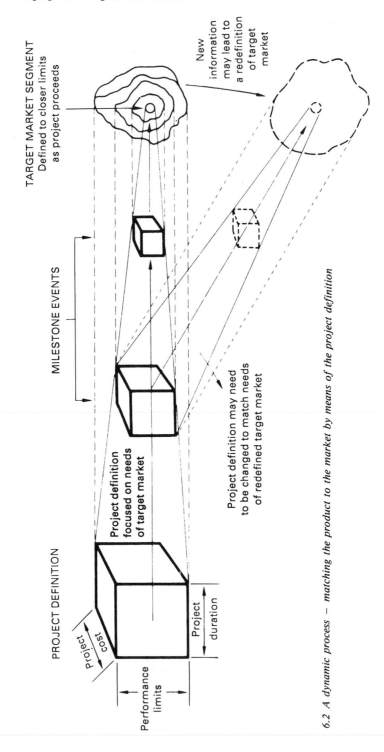

6.2 A dynamic process – matching the product to the market by means of the project definition

terms which do not unnecessarily constrain the freedom of the project team to formulate novel solutions. It establishes the criteria determining the planning objectives for all the tasks forming part of the project and, in particular, must lay down clear guidance on:

1. The technical performance required.
2. Cost limitations.
3. The project duration.

Portfolio planning

An R & D portfolio will contain a variety of projects, some large, some small, some approaching completion, others just started, and each calling for an allocation of scarce resources peculiar to its own characteristics. Many of these projects will fall by the wayside and be terminated before reaching the objective specified in the project definition. Thus the constituents of the portfolio will be changing continually, both in number and in respect of the demand they place upon the resources of the R & D department. The R & D director can never plan his portfolio and then turn his undivided attention to managing the fulfilment of the plan. Planning and replanning must be continuous. Yet the portfolio must have shape and be sufficiently stable for the working programme to be progressed smoothly.

The primary consideration for the portfolio must be the achievement of an output of new products or processes of the type required to further the corporate objectives of the organization at the time they are needed. This must be done within the budget allocation for that purpose. Since the budget covers expenditures on a wide variety of discrete and non-interchangeable resources, the portfolio must be planned in such a way that:

1. The available resources are used effectively.
2. The progress of projects is not hampered by shortage of resources.

Sometimes it may be found that the make-up of the portfolio is being restricted by the size of the budget allocation or the nature of the resources available. The R & D director must be on the constant lookout for evidence that this is occurring. Generally, he is unlikely to obtain an increase in his budget, although a well argued case, based upon a careful analysis of his optimum portfolio, may enable him to obtain a temporary increase in his financial allocation to tide him over a short-term problem. The detailed disposition of his funds is a matter, however, over which he has greater discretion. A study of the project proposals coming forward might suggest that an improved portfolio could be selected by an adjustment of his resources, perhaps by the disposal of underutilized equipment or staff, and their replacement by what is more appropriate to his needs. But this usually takes time, and for most practical purposes of portfolio balance it must be assumed that there is limited opportunity to match the resources to the desired portfolio. Consequently, the portfolio must be selected to make the most effective use of existing resources.

The number and size of projects in the portfolio

The number of projects a portfolio contains at any time depends upon two factors – the size of the projects, measured by the total resources needed to complete development; and their duration, largely determined by the rate at which the resources are allocated to them, i.e.

$$\text{Number of projects} \propto \frac{\text{Total R \& D budget for the period}}{\text{Average size of project} \div \text{Average} \atop \text{project duration}}$$

When balancing his portfolio, the R & D director must consider how many projects he wishes to be managing at a time, whether he should concentrate on a few projects or spread his resources over a greater number, and the rate at which they should be progressed.

A portfolio containing a high proportion of large projects is inherently more risky than one where the resources are devoted to a large number of small projects. If experience indicates that only 10% of all projects are completed successfully, then there is only a 10% chance of a profitable outcome from a one project portfolio. However, as the number of projects increases so does the probability that at least one of them will be successful. Moreover, within an acceptable overall risk level for the portfolio as a whole, it becomes possible to include projects of higher individual risk with the greater potential rewards which might be expected from one risky project.

Another advantage of smaller projects is that they can be fitted together more easily to match the available resources. A large project may absorb a disproportionate amount of a scarce resource. This may result in that project or others being paced by the available capacity of a particular resource, and may lead to severe underutilization of some other resources.

These arguments in favour of small projects must, however, be weighed against other factors. Small projects (i.e. small R & D investments) usually lead to new products with a small sales potential (i.e. total profit potential). Although the aim of maximizing the benefit:cost is a prime objective, the opportunities for finding products with a high potential sales volume and profit margin for a small development cost are obviously limited. Thus a portfolio of small projects can be expected to produce a steady stream of innovations most of which have a limited market potential. In general, this is likely to be undesirable in relation to the marketing department's product portfolio.

The ultimate success of any project depends as much on the quality of the project management as it does on the technical or marketing merits. But good management is a critical resource for most companies. Thus, when considering the number of projects to be included in a portfolio, it is not sufficient to evaluate them solely on their potential desirability. Some allowance must also be made for the quality of the management and its ability to realize this potential. We can see what may happen in practice when managerial ability is dispersed over a number of projects from the example in Table 6.1. In this case, two portfolios,

Table 6.1 *Effect of managerial dilution on portfolio performance*

	Portfolio A			Portfolio B		
	Cost £	Benefit £	Benefit cost	Cost £	Benefit £	Benefit cost
Project number 1	12,000	23,000	1.9	24,000	42,000	1.75
Project number 2	10,000	18,000	1.8	20,000	38,000	1.90
Project number 3	8,000	16,000	2.0	–	–	–
Project number 4	6,000	14,000	2.33	–	–	–
Project number 5	8,000	14,000	1.75	–	–	–
Estimated portfolio performance	44,000	85,000	1.93	44,000	80,000	1.82
10% increased R & D cost on Projects 3, 4 and 5	2,200					
10% decreased benefit from Projects 3, 4 and 5		4,400				
Likely performance adjusted for project management	46,000	80,600	1.72	44,000	80,000	1.82
5% increase in R & D cost due to dilution of supervision	2,200					
Likely portfolio performance	48,400	80,600	1.67	44,000	80,000	1.82

A with five projects and B with two projects, are compared. On the basis of the estimated benefit:cost ratios for the individual projects, A offers the higher potential overall benefit:cost (1.93 for A compared with 1.82 for B). But the project managers will be of varying ability. If one assumes that the most promising projects (i.e. Nos. 1 and 2) are given to the best managers, then project Nos. 3, 4 and 5 in Portfolio A will suffer some degradation in their performance through less competent management. For the purpose of this example it is assumed that this will result in a 10% increase in development cost and a 10% decrease in the benefit due to duration over-runs delaying launch. An additional loss in effectiveness is likely to result from the dispersion of senior management supervision over a large number of projects, limiting the attention that can be devoted to each.

Adjustment for managerial performance suggests that in practice the eventual outcome of Portfolio B may be more favourable than for A. Since both the portfolios in this example are small the penalties for managerial dilution have, perhaps, been overestimated. However, in the larger portfolios found in most laboratories, the effect of varying managerial ability on project performance can be very considerable.

The phasing of projects can also have a significant influence upon portfolio performance. There is a choice to be made of the rate at which resources are allocated to the projects selected for inclusion in the portfolio. They can, for example, be progressed as quickly as possible by concentrating effort on a few projects at a time, or alternatively, a greater number can be worked on for a longer period. Apart from the managerial advantages already discussed, there are additional benefits which might be derived from a concentration of effort. These can be seen by examining the simple example of a portfolio consisting of two projects possessing identical cash flow profiles; they can be worked on either in sequence or in parallel (Fig. 6.3). It is assumed that the products

(a) Parallel development

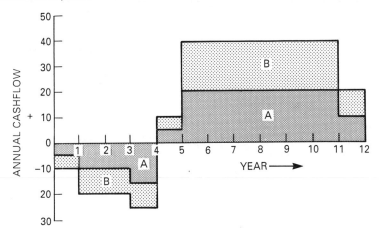

(b) Sequential development

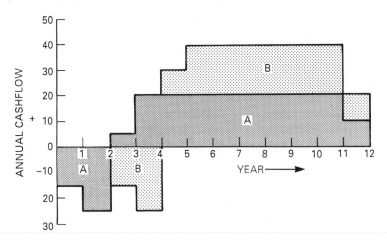

(c) *Comparative cash flow table*

| YEAR | PARALLEL DEVELOPMENT | | | | | SEQUENTIAL DEVELOPMENT | | | | |
| | Cost | | Benefit | | Net cash flow | Cost | | Benefit | | Net cash flow |
	A	B	A	B		A	B	A	B	
1	5	5			− 10	15				− 15
2	10	10			− 20	25				− 25
3	10	10			− 20		15	5		− 10
4	15	15			− 30		25	20		− 5
5			5	5	+ 10			20	5	+ 25
6			20	20	+ 40			20	20	+ 40
7			20	20	+ 40			20	20	+ 40
8			20	20	+ 40			20	20	+ 40
9			20	20	+ 40			20	20	+ 40
10			20	20	+ 40			20	20	+ 40
11			20	20	+ 40			20	20	+ 40
12			10	10	+ 20			10	10	+ 20
TOTAL	40	40	135	135	+ 190	40	40	175	135	+ 230

Note 1 Increase in net cash flow from + 190 to + 230 represents + 21% without taking into account the effect of discounting

2 If the product lives were shorter this effect would be even more marked. For example + 57% for a three-year reduction to a nine-year life

6.3 Financial benefits from project sequencing

developed will have a limited life curtailed at some date in the future determined by competitive pressures. It is possible to start both projects immediately and to complete development in four years. Alternatively, Project A can be completed before B is started, provided there are no technical reasons why the projects cannot each be completed in two years given the availability of the necessary resources. Examination of Fig. 6.3 shows that sequential development offers the following advantages:

1. Managerial effort is devoted to only one project at a time.
2. Any duration over-run on Project A can be accommodated simply by delaying the start of Project B without resource reallocations within the portfolio.
3. The innovation resulting from Project A is available two years earlier, thereby
   (a) giving it an increased product life of two years, and
   (b) possibly giving a commercial and financial benefit by being early on the market.
   (This is not reflected in the figures for the example.)
4. Project B is started two years later when improved technical and market

information should be available. Some of this advantage might be offset if R & D problems were encountered (due to the shorter development period there is less time available for resolving such difficulties).

5. The net cash flow position is considerably improved in years 3 and 4, although slightly worse in years 1 and 2.

6. The total positive cash flow and the benefit:cost of the two projects is improved. Discounting would further improve the comparison since the additional benefit is earned in the early years.

Although the example described is unrealistically simplified it does illustrate the considerable advantages to be gained by reducing the number of projects being worked on by a concentration of resources. These will frequently justify an increase in the total R & D expenditure in order to 'crash' a particular project.

Portfolio analysis techniques (see Chapter 2) are a valuable tool in ensuring that the size of the portfolio is not only minimized but also focused on areas of greatest potential. By this means ICI Pharmaceutical Division reduced its research targets from 60 to 30 in the period 1980–5.

In planning a portfolio we may conclude that it is generally desirable to limit the number of projects by:

1. Selecting large projects whenever possible, provided the number is sufficient to give an acceptable spread of risk and enables the efficient utilization of resources.

2. Concentrating effort at any time on only a few of the projects selected to ensure their speedy conclusion.

Importance *v* priority

The priority accorded to a project need not necessarily be related to its ultimate importance to the business. This arises because projects have different cash flow profiles. We can see how this can occur by examining the case illustrated in Table 6.2. In this example, both Projects A and B have been selected for inclusion in the portfolio, but because of shortage of resources only one can be started immediately; work on the other cannot commence for a year.

From this analysis it is clear that Project B with a net present value of £34,025 is much more *important* to the company than Project A with a value of £17,840. Yet because of its shorter life the loss from delaying Project A by one year £8,050 is greater than for Project B £5,440. Therefore, consideration of the different financial characteristics of the two projects indicates that *priority* should be given to Project A.

Project planning

Project planning and portfolio planning cannot be considered in isolation. The project plan should consist of a detailed programme of work specifying the

Table 6.2 *Effect of the cash flow profile on the priority accorded to a project*

PROJECT A			PROJECT B		

Project details

	PROJECT A			PROJECT B	
R & D cost	£5,000		R & D cost	£5,000	
Benefit	£10,000 per year		Benefit	£7,000 per year	
Project duration	1 year		Project duration	1 year	
Product life	From launch to year 4		Product life	From launch to year 16	

Net present value of project if started immediately (discount rate 15%)

Benefit			Benefit	
Year	Discount factors*	Net present value of earnings £	Discount factor for £1 received annually for 15 years at 15% = 5.575	£
2	0.87	8,700	Net present value of	
3	0.756	7,560	project benefit	
4	0.658	6,580	£7,000 × 5.575	= 39,025
Net present value of benefit		22,840	Less cost of project	= 5,000
Less cost of project		5,000	Net present value of	
Net present value of project		£17,840	project	= £34,025

Effect of delaying start of project by one year

Benefit			Benefit		
Year	Discount factors*	£	Reduction due to loss of 1st year's		
3	0.756	7,560	earnings	=	£7,000 × 0.87
4	0.658	6,580			£6,090
Net present value of benefit		14,140	Net present value of benefit	=	32,935
Cost			Cost		
Year	Discount factors*		Year	Discount factor	
2	0.87	4,350	2	0.87	4,350
Net present value of project		£9,790	Net present value of project		£28,585

Reduction in net present value by delaying start by one year

	£		£
Net present value if started immediately	17,840	Net present value if started immediately	34,025
Net present value if delayed one year	9,790	Net present value if delayed one year	28,585
Reduction in net present value by delayed start	£8,050	Reduction in net present value by delayed start	£5,440

* For ease of computation the *discount factor* used has assumed that all costs and all benefits were incurred at the beginning of the year.

budgeted cost for each task, the resources needed, and the timescales. These have to be integrated with the requirements of the other projects in the portfolio to ensure that the total demands for each type of resource are kept within the level that is available. Usually, difficulties will be experienced and it will be necessary to change the intended project plans in order to smooth the total demands upon the resources. Sometimes, this can be achieved by rephasing the work schedule without affecting the planned completion date, but frequently it is found that the duration of the project must be extended if serious interference between projects is to be avoided. In Chapter 5 it was shown that, in certain circumstances, linear programming can be used to aid this task. However, because of the problems which the projects will experience during their execution and the difficulty of forecasting the detailed requirements with any degree of accuracy, all that can be expected at an early planning stage is the identification of serious mismatches which the plans must attempt to avoid.

Many projects comprise a large number of smaller projects or tasks, some of which may be the responsibility of outside suppliers such as contract R & D organizations or the designers of other parts of the total system. At project initiation it must be decided:

1. Who is responsible for which task or constituent part of the total system.
2. What is the critical interface information which determines the output from one part which is the input to another.
3. The timescale for tasks which are sequentially dependent.
4. The method for monitoring, coordinating and controlling progress to resolve problems arising in relation to technical achievement, cost and duration performance.

Von Hippel [1] has shown that the initial decisions on partitioning the tasks for a programme both within the company and between suppliers can have an important impact on the project efficiency and effectiveness. Once the partitioning has been completed the resource allocations can be assessed.

When the outlines of the resource allocations to the project are agreed its detailed planning can proceed. Since there is every likelihood that the original plan will have to be modified often several times during the project's life, provision must be made for periodic reviews. These reviews are critical events in project control. Thus it is necessary to identify the 'milestone' events within the programme and schedule the programme of work so that the information necessary for the review is available at that time. The plan should also ensure that the minimum of resources are devoted to work not essential for the review for this could result in wasted effort if the review led to a decision to terminate the project.

Although major re-evaluation normally takes place at the milestone events, some provision must also be made for major unforeseen difficulties. There must be some mechanism to ensure that, when things are going badly astray between

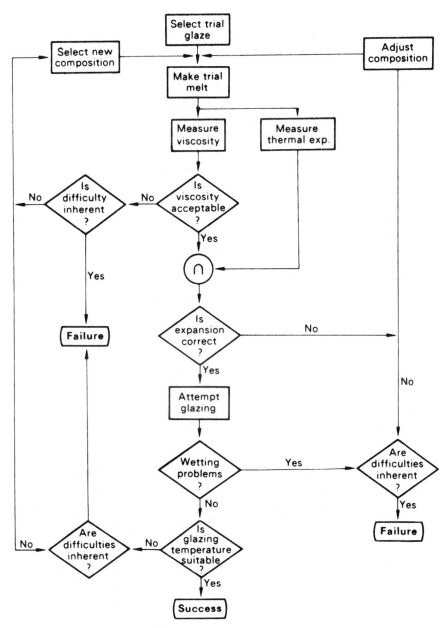

6.4 Example of a research planning diagram

Source: D.G.S. Davies, 'Research Planning Diagrams', *R & D Management* Vol 1, No 1, Oct 1970

milestone events, additional resources are not allocated to a project without a re-evaluation. We shall see later that the control information can be used for this purpose.

A comprehensive plan for a project from start to finish can only be indicative of how it will proceed in practice. Many unanticipated problems will cause deviations from the initial plan, but since it is the completion of the project which is of value to the business this objective must never be lost sight of. The consequences of all divergences from the plan must always be examined in relation to the final objective. Nevertheless, most attention will be directed towards the conclusion of the work leading to the next milestone event.

When planning a well established activity, such as building a ship or a power station, it is possible to analyse every activity and make reasonably accurate estimates for the costs and time needed to complete it. These can be combined within an overall plan for the whole project. Readers will be familiar with the now widely used network planning techniques such as PERT which provide an effective planning and control tool for this type of operation. Although these techniques can be used successfully during the later stages of R & D programmes, they are of limited use when there are major uncertainties regarding what activities will be required.

Thus a much more flexible planning system is required for the early stages of a project. The system must be able to take account of the uncertainties and indicate the action to be taken in the event of any part of the programme failing to achieve its objective. Contingency planning must be built into the overall plan from the very beginning. One technique for doing this has been devised by Davies [2,3]. His 'research planning diagrams' (Fig. 6.4), drawing upon the methods of both network planning and computer programming flow charts, enable a full analysis of alternative activities, possible outcomes and the resulting actions which should be taken. Subjective probabilities can be applied to the component links in the diagram, thereby permitting quantitative analyses of the cost and time probabilities to be made.

It must be stressed, however, that the main purpose of the plan is to enable effective project control. It provides a basis for the evaluation of progress by identifying the critical elements of the programme and expressing them in terms of effort and time. A project deviation from the plan is a signal for action. This may result in either a reallocation of resources to recover the timescale originally planned or a reformulation of the plan to reflect the new realities.

The plan can, however, be only as good as the information upon which it is based. Acceptance of the need for planning flexibility should not be an excuse for poor estimating performance. Improved planning estimates lead to more realistic plans and consequently better project control. If the planning and control systems are taken seriously, greater attention is likely to be paid to estimating accuracy. Thus we have a closed-loop system (Fig. 6.5).

A - Effective planning and control

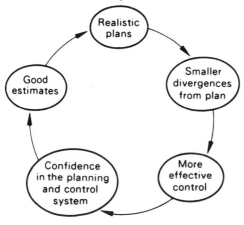

B - Inneffective planning and control

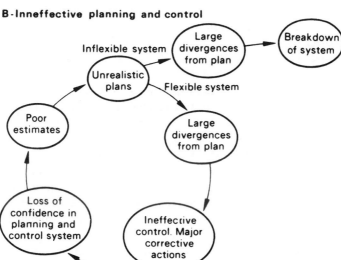

6.5 Estimating accuracy, planning, and control effectiveness as a closed loop

The main features of any project plan can be summarized as follows:

1. The allocation of resources in relation to the portfolio.
2. Definition of the programme of work in terms of activities, resources and time.
3. Identification of critical or 'milestone' events.
4. Identification of critical tasks.
5. A schedule of work related to the milestone events.
6. The integration of all the activities within an overall plan, using the planning techniques most appropriate to the stage of the project.

Project control

Information

Accurate and timely information is the life-blood of effective project management. All control systems are based upon the comparison of information about the situation as it is, with a predetermined standard of performance which the plan calls for. When there is a variance between the measured achievement and the standard, as is usually the case, action is required. Desirably the action should be designed to bring the project back onto course, perhaps by a reallocation of resources or priorities. This is not, however, always possible if major difficulties are experienced; in these circumstances it may be necessary to change the standards, i.e. revise the plan.

The standards which provide the basis of R & D managerial control are:

1. Project evaluation criteria and the estimates and assumptions upon which the project selection decision was based.
2. The project definition.
3. The project plan.

Throughout the duration of a project it is necessary to ensure that the information provided to management is adequate for a review of its performance in respect of all three standards. The attention of project management naturally focuses upon the achievement of the project plan. But this is too parochial an approach. A project might, for example, be proceeding to plan with every expectation of achieving the objectives set by the project definition. But if some factor external to the R & D department changes, the project may cease to be commercially viable, irrespective of its technical progress. In these circumstances, it may be necessary to modify the project definition or even terminate the project in the light of this additional information.

Effective control is impossible without adequately documented standards. It is essential to ensure that this information is always available, as its absence is one of the most frequent causes of poor control. A project is selected for what may appear good reasons at the time, but if the assumptions are not recorded, it becomes difficult to determine whether a particular item of new information has a significant implication for the project. In the longer term periodic re-evaluations lead to a complete review of the project using all the information available at that time. But for maximum effectiveness a system is needed whereby the significance of a new piece of information is noted immediately and, when necessary, changes in the plan introduced without delay. This enables continuous control, rather than the intermittent control which results if modifications are only introduced at infrequent reviews.

We have seen already that the control information arises partly within the R & D department and partly outside it. The external sources of information are those where deficiencies are most likely to occur. This occurs because an adequate communication system between different parts of the organization is often lacking. Continued liaison and cross-checking of information with the marketing, production and finance departments is necessary throughout the project. This is particularly important for long duration projects where the basis for commercial viability may change considerably during development. Yet it is often found that liaison between departments is poorest where technical gestation periods are long. Thus the product which eventually emerges from the laboratory meets the needs as forecasted when the project was started. But the market may have changed and the product is no longer required in its original form. Although the marketing department was well aware of this change it may have neglected to communicate this information promptly to the R & D department which continued to spend money to no good purpose.

The company's organizational structure has an important bearing upon the flow of information across departmental boundaries. This is more likely to occur naturally when the project manager is vested with a high degree of commercial responsibility as in 'venture' or 'matrix' management (see Chapter 7). In other forms of organizational structure, where the project manager's attention is concentrated almost entirely upon the technical progress of the project, the research director must ensure that he or another member of senior R & D management is personally responsible for this aspect of the communication system. It must not be neglected.

For the purpose of controlling the technical progress of the project, it is necessary to have up-to-date information on the costs incurred, the time elapsed and the proportion of work completed. Although the company's accounting system can provide some of the information required, it will usually be left to R & D management to establish a system for determining the cost data for individual projects. Since R & D is a labour intensive activity the major element is staff cost. This presents no problem where a project team is employed exclusively on the one project. Where this is not the case it is necessary to institute a system for collecting information on how staff time has been allocated between the projects they are working on. This means that they must be required to complete time sheets. Although this is often unpopular with professional staff it is unavoidable if meaningful control information is to be obtained. It is also necessary to collect data on other items directly attributable to the project such as materials used and the expense of laboratory and test facilities. The overhead costs of the R & D department as a whole can be allocated to individual projects as a percentage of their staff costs. Since staff account for the majority of the costs this basis for the allocation of overhead is reasonably accurate. The allocation of overhead costs to projects does not, however, have any real meaning for project control which is concerned with the management of resources directly employed on the project. There is, therefore, no purpose in including overheads in the control calculation.

Information on the technical progress of a project is more difficult to obtain and involves an element of subjective judgment. Usually, however, it is not so difficult as might be expected. The project will have been divided into a number of well defined tasks within the plan. At any time, some of these will have been completed and for others, which will be progressing well, it will be possible to estimate with reasonable accuracy the proportion of the work accomplished. There are likely, however, to be a few tasks which have experienced problems as yet unresolved and an estimate of their progress to date must remain a matter of judgment; however, unless these form a major part of the project they will not introduce a high degree of inaccuracy into the estimates (Table 6.3).

Table 6.3 *Estimation of project progress*

Task	Estimated cost (£)	Task cost as % of total cost	Cost to date (£)	Percentage completion	Comments
A	10,000	20	10,500	100	
B	2,000	4	1,600	100	
C	7,000	14	5,000	70	
D	5,000	10	1,500	25	
E	6,000	12	5,000	10	New estimate to completion £6,500
F	5,500	11	–	–	
G	6,500	13	4,500	65	
H	5,000	10	1,000	10	
J	3,000	6	–	–	
Total	£50,000	100	£29,100		

% Project completed $= 20 + 4 + 9.8 + 1.2 + 0 + 8.5 + 1 + 0$
$= 47$

% Estimated cost
spent to date $\quad = 58.2$

The use of control information

Planning and control should be parts of a unified system. The plan indicates how the project is expected to progress, subsequent information provides a measure of how well the performance of the project is matching the plan, and the control function compares the two and leads to managerial decisions and actions for the future conduct of the project. Thus the techniques used in the initial planning of the project also provide the basis for managerial control.

The planning and control techniques may, however, be changed as the project progresses. In the early planning stages it might be considered that the uncertainties are such that sophisticated techniques are unsuitable. However, as the details of the project clarify it becomes possible to use them. Thus a simple bar chart may be considered appropriate for initial planning, only to be replaced by network techniques of increasing refinement as the tasks and their relevant costs and durations become more clearly defined. The practice of adopting a standard planning and control technique throughout an R & D department should, therefore, be questioned. Although there are obvious advantages in using one technique, these have to be balanced against the improved control achieved by using a variety of techniques according to their appropriateness to the requirements of individual projects.

Management requires for its control system the ability to:

1. Assess the progress of each task in terms of cost and duration.
2. Identify those tasks which are falling behind schedule and assess the likely effect they will have on the overall progress of the project, i.e. delays in the critical activities of the network.
3. Measure the progress of the project as a whole in relation to the planned cost and completion date.

The first two items should present little difficulty if an adequate planning system has been installed. All that is necessary is to compare performance with plan for each individual task. When a discrepancy is noted, the manager must decide if it can be rectified within the existing resource allocation, whether additional resources are needed or, as a last resort, whether it necessitates a revision of the plan. The method of presentation of the information should automatically draw his attention to those parts of the programme requiring managerial action.

He also needs to monitor the overall progress of the project. In doing this, he must consider all the information available to him in an easily digestible form. Some parts of the project will be ahead of plan, some behind, some will have been completed at a greater and some at a lower cost than estimated. These, however, can all be represented by an assessment of the overall technical progress and the cumulative expenditures against time. Although the actual figures of performance against plan at a given point in time are useful, the trends over a period are likely to give a much better indication of how the project is progressing. This data is best presented in the form of graphs of (a) cumulative cost *v* time, (b) technical progress (as calculated in Table 6.3) *v* cost, and (c) technical progress *v* time (Fig. 6.6).

It is useful to draw all three of these graphs since they are likely to give different insights into the performance of the project. This can be seen from an examination of the two projects plotted in Fig. 6.6. Project 1 is seen in Fig. 6.6(a) to be absorbing costs at a greater rate than planned. This might be interpreted as a danger signal. Reference to Fig. 6.6(b) however, shows that the high rate of expenditure was accompanied initially by a rate of technical progress commensurate with the cost. Later a technical problem was encountered which slowed progress and the costs increased beyond those planned for the progress achieved. In spite of this, Fig. 6.6(c) suggests that the planned completion date might still be achieved provided the problem causing the current difficulty is overcome quickly. Thus it can be seen that if the profitability of this project is more sensitive to programme slippage than to development cost its future is not under immediate threat. Project 2 shows a quite different pattern.

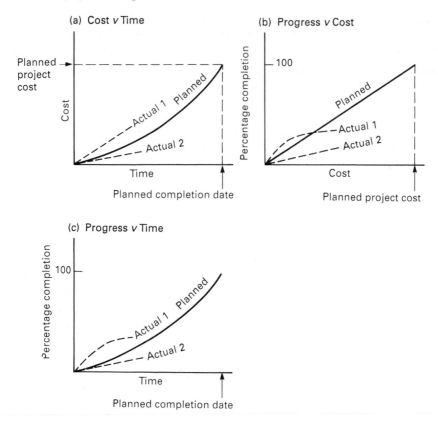

6.6 *Project control charts*

It has progressed slowly from the start. Although the expenditure against time is below that planned, technical progress has been slow, both in relation to cost and time. Whether the cause of this is poor management or technical difficulties, it appears to be a candidate for termination. The constant trend for all three graphs since initiation suggests that the problems associated with this project have been with it from the beginning and earlier corrective action could probably have been taken.

Figure 6.7 shows a simple control chart used by the British government to assess progress on major projects. It is based solely upon financial data – actual costs and estimates.

Resource allocation and management

The difficulties associated with the efficient allocation of resources have been a continuing theme throughout this chapter. This arises because:

1. The total resources available to R & D management need to be relatively

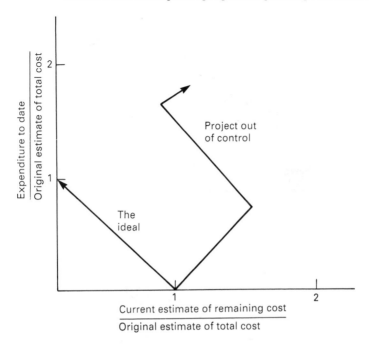

6.7 *Project control chart based on financial data*

stable over time. Since the availability of staff is the major item determining the rate of expenditure frequent adjustments cannot be made without affecting the level of staffing. Such a policy is likely to bring about a rapid fall, both in morale and productivity.
2. The resources available are invested either in items of equipment which are a fixed cost whether or not they are used or in staff who possess specific and non-interchangeable capabilities.
3. Each project absorbs a different combination of these resources. Not only do the requirements change, but it is not possible to forecast precisely what they are likely to be at some date in the future.

This problem is represented in diagrammatic form in Fig. 6.8. When the number of projects and the types of resource increase, the problem of effective programming can become exceedingly complex. Inevitably, accurate resource management is impossible. As a consequence, some projects will suffer delays due to a shortage of critical resources while at the same time other resources will be under-utilized.

It would be possible to adopt a mechanistic approach to resource allocation were it not for the difficulty of planning the requirements accurately. Production planning departments in a similar situation are increasingly availing themselves of the scheduling flexibility provided by the computer, but this is only possible because the individual requirements can be assessed with some precision. In

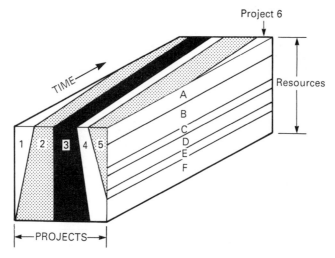

6.8 *Balancing the budget – problem of resource allocation*

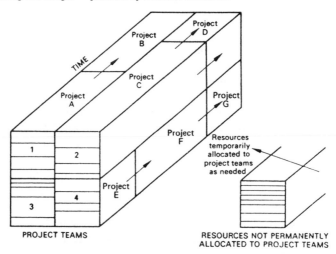

6.9 *Resources allocated primarily to project teams*

R & D less sophisticated solutions must be sought.

Another complicating factor that arises with a simple mechanistic approach is the extent to which the success of a project depends upon people. Whereas a change in the scheduling of a job on to a piece of production equipment affects the timing but not the duration of the task, this is not the case when people are involved. Project delays occasioned by a shortage of staff or resources have a disproportionate effect upon the morale of the project team and their determination to achieve the planned dates. This can produce a lack of urgency causing the project to slip even further behind schedule. But the reverse is also true. When the team accepts a plan as realistic and is not unduly hampered in its accomplishment by influences outside their control they are likely to rise to the challenge of the problems they encounter and will do their utmost to achieve their targets.

A central consideration in the management of resources must, therefore, be the morale of the project teams. Thus a project-centred approach is likely to achieve a greater overall effectiveness of the R & D effort than one where the emphasis is placed solely upon the efficient allocation of resources. Admittedly, some under-utilization of equipment or staff may result, and may even impose a limitation on the size of the portfolio; but this can be a small price to pay for the added effectiveness of the project teams. For the efficient allocation of resources does not guarantee they will be used to good purpose.

Authority to allocate and control resources is one of the main determinants of a person's power within an organization. So consideration of the system of resource management cannot be separated from the organizational structure of the firm. Figure 6.9 shows in a simplified form how this occurs when the project team is made the basic unit for resource allocation. The project team is formed on a semi-permanent basis, each team possessing a nucleus of staff and any items of equipment which can be allocated economically for their exclusive use. Delays in the completion of a project are accommodated by putting back the date for the commencement of the next project. Some resources cannot be given permanently to project teams. These, which are controlled centrally, include major items of costly equipment, staff who provide a service to all project teams, and additional technical staff who can be attached on a temporary basis to meet the short-term requirements of a particular team.

A system based upon the project team cannot, of course, remove the inherent problems of resource management. It can, however, alleviate them in a number of ways:

1. The overall problem is disaggregated into more easily manageable parts. Thus, many of the resources are removed from day-to-day consideration since they have been allocated to project teams semi-permanently.
2. The members of a project team have the assurance that they can work towards their objectives without interference to the size and membership of the team.
3. The attention of senior management is focused on the allocation of those resources likely to cause critical problems, i.e. those which are allocated centrally.

These advantages must be weighed against the constraints they impose on senior R & D management:

1. There is a loss of detailed control by senior management of the work being done by many members of their staff. Only in exceptional circumstances should members of a team be transferred. The whole rationale of the system is negated if the position of the project leader as controller of his own resources is undermined. This has important implications for both the

organizational structure and the attitude and leadership style of senior management.

2. The number of projects in the portfolio must remain reasonably constant over a period of time, since it is difficult to increase the number of project teams without withdrawing members of existing teams and thereby sabotaging the system. But as we have seen already, it is usually desirable to restrict the size of the portfolio. Thus any disincentive to project proliferation may be considered to be not entirely undesirable.

3. There will be periods when there is insufficient work for some members of the project team. There is, of course, no reason in these circumstances why they should not be transferred temporarily to assist in other work, provided the discretion rests firmly with the project leader. The membership of the team may, in any case, have to be adjusted from time to time to meet the changing needs as a project progresses.

4. Success depends upon the ability of senior management to control effectively the resources they retain direct control over.

It is not possible to prescribe one system as being more appropriate than another because of the varying nature of the problems encountered in different industries and companies. As a generalization, however, it can be said that the more complex the resource management problem becomes, the greater the advantages to be derived from a project-centred system. There is, however, certainly little point in it when the laboratory is so small that the R & D director can handle the matter personally without difficulty, because he possesses an adequate knowledge of the implications of a decision for the projects affected.

Staff motivation

The motivation of R & D staff is a topic which has received considerable attention by behavioural scientists. It is beyond the scope of this book to examine this subject in any great detail although some aspects which have important implications for the organization of the company will be examined briefly in Chapter 7.

The existence of a plan and a formal control system does not in itself ensure that the project objectives will be achieved. Staff must be motivated and decisions taken in the light of changed circumstances. Figure 6.10 illustrates the main elements involved in the 'planning' and the 'doing' of a project.

So far in this chapter we have been concerned with systems of control for project portfolios and resources. There is an implicit assumption that logical analysis leading to the establishment of realistic objectives and plans, followed by the analysis of control information and appropriate corrective action, should result in effective management. But the art of management is making things happen. And in R & D, more than in any other aspect of industrial management, success depends upon people. Creativity and entrepreneurship, so important in R & D, cannot be planned, but the conditions in which they can flourish do depend upon managerial decisions. Thus planning and control only provide a framework within which people work. The systems may direct the effort but the energy devoted to their realization is largely dependent upon the motivation of the people involved.

Human attitudes impinge on every aspect of the planning and control system. The plan itself depends upon the accuracy of the estimates upon which it is based. But these estimates are highly subjective, and rely upon the judgment of individuals who may not always take them seriously. The implementation of the plan can only be effective if it is perceived to be realistic by those who are responsible for its achievement. If they reject it, albeit subconsciously, they will work towards their own targets poorly formulated and uncoordinated. Management often deludes itself by the apparent comprehensiveness of its plans only to receive a rude awakening when the outcome bears little relationship to it. The plan may have been adequate, but it failed because the implementers thought otherwise. Whether they were right or wrong in so thinking is largely irrelevant.

A similar situation may arise when the plan is changed as a result of action based upon control information. Any consequent reallocation of resources must

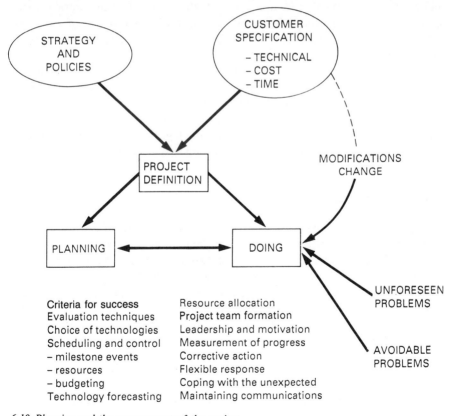

6.10 Planning and the management of the project

be believed to be based on sound reasoning if the necessary acceptance and cooperation is to be forthcoming. Otherwise, management control will be interpreted as an interference and a source of frustration. The changes will be rejected, not overtly, but the effect will be equally serious and will manifest itself in poor morale and a reluctance to commit that extra effort so vital to the energetic conduct of R & D projects.

Identification is the key to success. The importance attached to the involvement of the R & D director in corporate decisions, and his identification with corporate objectives and strategies, was stressed in an earlier chapter. It is equally important that the R & D director takes the views of his staff into account when formulating policies and plans. It is a human failing to deny one's subordinates the same degree of involvement that one expects from one's own superiors.

Thus the character and leadership style of senior management is a vital ingredient in ensuring the cooperation of all concerned. A participative style would appear to be indicated.

It must be accepted, however, that no rules for universal application can be made. There are many examples of laboratories which appear to be 'poorly managed' in a technical sense and yet are highly successful. In contrast, there are others which consistently fail to produce useful innovations in spite of their use of sophisticated techniques. It is dangerous to judge on the basis of superficial measures.

One of the most powerful motivators is a sense of purpose. This can be engendered by a formal system of planning and control. But the leadership of an enlightened manager may be equally effective. The advantage of a formal system is that it provides a rational basis for the setting of project objectives, the planning of detailed work, and decision-making. However, it will fail if used as a procedure for allocating blame, rather than as a means of identifying where support is needed.

Managing the cash flow curve

A typical cumulative cash flow curve from the initiation of a project to the eventual withdrawal of the product from the market is illustrated in Fig. 4.7. Because of the problems of estimation discussed in Chapter 5 it is not possible to quantify the shape of the curve accurately in advance. Nevertheless it is possible to establish the general shape of the curve from past experience with similar projects. This should be done and is likely to provide valuable insights and guidance for managerial decisions. It can be seen that the curve approximates to three phases of a project:

1. *Applied research to feasibility.* During this phase the curve is flat indicating a relatively low rate of expenditure.
2. *Product development.* This is a period when the negative cash flow rises rapidly. Prototype or pilot plant testing are major contributors to this.

3. *Project completion.* As the project nears completion the curve tends to flatten in the period immediately before product launch. This is usually due to final preparation activities which may be time-consuming but cost little. Almost invariably this time could be shortened by effective management action.

Typically the cost to the end of the first phase is 10% or less of that in the second, a fact many managers seriously underestimate. This is one explanation why so many new companies fail before a potentially successful product is launched.

We have seen in Chapter 5 that it is impossible to estimate the shape of the cash flow curve with any degree of accuracy. Nevertheless it must be recognized that the actual shape for any project depends to a great extent upon R & D management decisions. Because of the characteristic slope of the curve the reader can see that:

1. The date at which the product is withdrawn from the market is largely a function of its specification determined early in the life of the R & D project.
2. The period for which a product has a useful life is therefore almost entirely a function of its launch date. R & D activity should be geared to the timely completion of projects through a concentration of resources and a sense of urgency.
3. Time can be purchased most cheaply when the costs are low. Thus a sense of urgency should be introduced early in the programme. As the project progresses it becomes increasingly expensive to recover slipping schedules.

It must be recognized that the curve is a representation of the consequence of all the managerial decisions which are taken during the progress of the project. Thus it is possible to modify its shape. In general the aim must be to bring forward the date of launch and to allocate resources in a way that will achieve this objective, as described in the next section.

Table 6.4 *Project evolution and its control impliction*

	Feasibility	→	*Product launch*
Uncertainty	High	→	Low
Definition of market need	Broad	→	Precise
Need for creativity	High	→	Low
Ability to modify technical specification	High	→	Very low
Accuracy of estimates	Low	→	High
Cost	Low	→	High
Money at risk	Small	→	Large
Management style	Participative (essential)	→	Participative (desirable)
Frequency of contact with other functions	Low	→	High
Suitability of sophisticated management control techniques	Low	→	High

The evolution of a project as it moves from initiation to the market launch of the product exhibits a change in the quality of the information available for decision-making and, consequently, the type of control techniques which are appropriate. Sensitivity to these changing circumstances (Table 6.4) is the hallmark of the good project manager. In some cases it may be found necessary to change the project manager as the project progresses.

The shape of the cumulative cash flow curve shows clearly the need to decide at the earliest possible date whether or not to terminate a project. Since a high proportion of the projects which are initiated are cancelled before they result in marketable products (Fig. 6.11) any delay in making decisions to terminate can result in a disproportionate increase in cost.

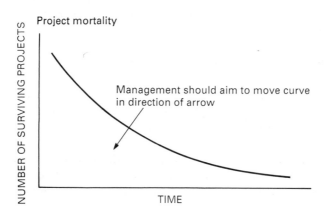

6.11 Project termination – the need to decide early

Control priorities – the need for urgency

In an increasingly competitive world it is becoming more essential than ever to shorten development times. This is not only true of new technologies where technical performance is always the key to success but also of the mature industries where the advances are more incremental. Furthermore, where product lives are shortening any delay in the launch of a new product can have a significant effect upon its useful market life and consequently its total sales volume and profitability. Allied to this, teething problems must be eliminated. No longer is it possible to regain lost markets by rectifying early problems in service; the time is too short. Thus it is essential to 'get it right first time'. The satisfaction of these conditions depends upon the timely completion of R & D, good detail design, rigorous testing, and attention focused on all aspects of quality at every stage of the innovation process.

We have seen that the three variables to be taken into account in planning and controlling a project are its technical performance, the development cost and duration. Between these there are tradeoffs and it is essential that emphasis be placed upon that variable which brings the greatest corporate benefit. The ideal of the highest performance product at an early completion date and at minimum project cost is impossible to achieve in practice. Choices have to be made.

The technical performance must be appropriate to satisfy the market need. If it is too low it will not be competitive. If too high, an understandable motivation for the technologist, it will add little competitive advantage but will incur cost and time penalties.

Most critical is the tradeoff between development cost and duration. This will vary with the nature of the project but, in general, an early launch date yields a greater benefit than the costs of achieving it. Beswick [4] has calculated that for a typical electronic product the profit sensitivities are:

1. 50% development cost over-run produces a 3.5% profit loss.
2. 9% increase in product cost produces a 22% profit loss.
3. 6 months slippage incurs a 33% profit loss.

This may be an exceptional example for a fast-moving industry, but it does highlight the need to undertake a sensitivity analysis and concentrate effort where it will yield the greatest return. Frequently, it is found that the benefits of early product launch justify the allocation of additional R & D resources in order to shorten the development time. However, it must be recognized that it may be difficult to obtain sanction for this course of action where R & D budgets are tight and in the absence of a well-argued case supported by sensitivity analyses.

There are a variety of ways to shorten the development time. Examination of the cumulative cash flow diagram suggests that time can be bought most cheaply at the two ends where the curve is flattest. Amongst the managerial options are:

1. Inculcating a sense of urgency from the outset of the project.
2. Parallel approaches in applied research (see Fig. 5.8).
3. Developing production facilities in parallel with product development; this may involve costly modifications later in the programme but is a risk an increasing number of companies are prepared to accept.
4. Focusing effort by reducing the portfolio size.
5. Maintaining the tempo until the date of launch, for example by not appointing the experienced manager to a new project prematurely.

One example of the application of these principles is provided by Glaxo which reduced the development time for the Zantac anti-ulcer drug to under 5.5 years from first synthesis to market launch compared with the industry norm of 7 to 10 years. Many activities were run in parallel. These included R & D stages, toxicity and clinical trials wherever possible, and commencing the construction of the production facility three years before the drug received approval. Volvo has achieved reductions of the order of 25% in the development time of new trucks by improving its decision-making processes, ordering production presses before design is completed and running work in different functions in parallel.

The contribution made by good management of the total innovation process as discussed in this book must not be neglected. Gold [5], for example, in his research into this aspect of important technological advances in Japan, USA and Europe identified the following contributory factors:

1. Reliance on external resources:
   (a) buying or licensing advances to catch up;
   (b) buying or licensing to gain competitive advantage;
   (c) buying firms with desired technological advances;
   (d) contracting for external development.
2. Reliance on intensified internal programmes:
   (a) increasing rewards for successful performance;
   (b) organizing internal competition in research;
   (c) initiating simultaneous R & D on successive stages of innovation.
3. Reliance on innovative R & D management strategies:
   (a) peer review to accelerate progress;
   (b) transfer responsibility to accelerate progress;
   (c) closer integration of R & D with other functions.
4. More basic issues:
   (a) balancing R & D project portfolios;
   (b) organizing and evaluating R & D programmes.

Summary

The uncertainties inseparable from R & D are nowhere more evident than in project planning and control. This does not reduce the need for planning – rather the reverse – but it does mean that it must be both realistic and flexible if it is to be effective.

The main elements of a planning and control system were identified and discussed. These were: the setting of project definition and objectives; a plan for achieving the objectives; a means for comparing actual with planned performance; and managerial action.

As a project advances from applied research to final product development it undergoes a gradual transformation. This must be reflected in the control system. The market need also becomes more clearly defined and the major uncertainties surrounding the decision-making information are resolved. At the same time there is a rapid increase in the rate of expenditure on the project. Thus the growing importance of making the right decisions is matched by the improved quality of the information needed for decision-making. Management must be sensitive to these changes. An understanding of this process, while essential for R & D management, is of importance also to financial management since it affects the reliance they can place on the data they are given and to marketing management who are an important source of much of the information required. R & D management must adapt its control systems to meet this changing situation and may also find it necessary to modify their managerial style.

In circumstances where a high proportion of projects are prematurely terminated, it is essential that the control system enables timely decisions in view of the rapid escalation of expenditures. Thus one of the main considerations must be the design of a system which facilitates continuous, rather than intermittent, control. The project definition was seen to play an important part in defining the purpose and objectives of a project. It should be formulated with as clear an understanding of market needs as is possible at the time. While precise in terms of objectives, the project definition should avoid inhibiting novel solutions by stating how their objectives should be achieved.

It is useful to draw a clear distinction between the project definition and the project plan. The plan which details the work to be undertaken with cost and duration targets should follow an exhaustive search for alternative solutions to the problem.

The information used in initial project selection is also a useful reference standard. The recording of the data and assumptions used in selection enable the comparison with information gained later. Variations revealed and their implications for the success of the project provide a useful input to the control system.

Portfolio planning was also discussed. It was suggested that there was a strong argument in favour of limiting the number of projects in a portfolio, provided there was an acceptable spread of risk. A distinction was drawn between importance and priority. The most important project was not necessarily that which should be given priority within the work schedule.

Resource management, within constraints imposed by the financial budget and the availability of human skills and equipment, provides the R & D director with a demanding task. Resources fixed in size and nature have to be allocated to projects whose demands vary with time and are difficult to forecast, with the consequence that some degree of inefficiency is inevitable. It was suggested that the maintenance of the morale of the project teams is often more important than extracting the maximum efficiency in the allocation of resources. A project-centred approach, where fixed allocations of staff and some items of equipment are made to project teams on a semi-permanent basis, has advantages. Additional needs are satisfied from a centrally held pool of resources allocated by the R & D director or a senior member of his staff. In situations where resource management is a complex problem, the gains in project team effectiveness are likely to outweigh minor inefficiencies in the allocation of resources.

The motivation and morale of staff was discussed again in the final section of the chapter. Planning and control systems do not ensure effectiveness by themselves. They provide a valuable structure for project decision-making but only when accepted by the staff who have to operate them. Unrealistic plans and inflexible control destroy their own credibility and may, in some cases, lead to loss of confidence in the management with disastrous consequences for the projects concerned.

References

1. Von Hippel, E. 'Task Partitioning: An Innovation Process Variable', *Research Policy,* **19**, No. 5, Oct. 1990.
2. Davies, D. G. S. 'Research Planning Diagrams', *R & D Management,* **1**, No. 1, Oct. 1970.
3. Davies, D. G. S. 'R & D Tactics: Applications of RPD Decision Analysis', *R & D Management,* **12**, No. 2, Apr. 1982.
4. Beswick, R. quoted by Charlish, G. in: 'Designs on a Growing Market', *The Financial Times,* 4 Oct. 1985.
5. Gold, B. 'Approaches to Accelerating Product and Process Developments', *Journal of Product Innovation Management,* **4**, No. 2, June 1987.

Additional references

Batty, J. *Accounting for Research and Development,* Business Books, 1976.

Bridges, R. C. and Martin, T. J. ' "APPLE": A Dynamic Financial Simulation of R & D in a Technologically Based Business', *R & D Management,* **8**, No. 2, Feb. 1978.

Dougherty, D. M. *et al.* 'The Lasting Qualities of PERT: Preferences and Perceptions of R & D Project Managers', *R & D Management,* **14**, No. 1, Jan. 1984.

Dunne, E. J. 'How Six Management Techniques are Used', *Research Management,* **XXVI**, No. 2, Mar.–Apr. 1983.

Parker, R. C. *Going for Growth: Technological Innovation in Manufacturing Industries,* Wiley, 1985.

Pearson, A. W. 'The Management of Research and Development', in *Technology and Management,* Wild, R. (ed.), Cassell, 1990.

Roman, D. D. *Science, Technology and Innovation: A Systems Approach,* Gird Publishing, 1980.

Rosenau, M. D. 'Faster New Product Development', *Journal of Product Innovation Management,* **5**, No. 2, 1988.

Tyman, W. G. and Lovelace, R. F. 'A Taxonomy of R & D Control Models', *R & D Management,* **16**, No. 3, July 1986.

White, P. A. F. *Effective Management of Research and Development,* Macmillan, 1976.

7

Organization for innovation

The successful firm of the future will be one which is structured so both external and internal problems are given appropriate and continuous attention. Beyond this, the management structure will be conducive to innovation. The search for opportunities and problems will be institutionalized and continuous, the internal productive cycle will be R & D oriented, and manufacturing and marketing will be flexible and responsive to change in product-market mix.

H. Igor Ansoff

Introduction

In this chapter, we shall consider the organizational implication of the factors we have identified as important for the process of technological innovation. All companies engaging in R & D have, at some time or other, experimented with changes in their organization, in an attempt to create the right working environment for effective innovation. After a brief honeymoon period, the new organization is always found to exhibit a number of shortcomings. As we examine the criteria to be considered, we shall see why this is so, and why any organizational solution will at best be only a compromise. Theoretical considerations are supported by studies of business organizations which provide valuable information on organizational patterns companies have adopted, and the resultant effect upon managerial behaviour and effectiveness in innovation. Amongst the better known of these researches is the work of Burns and Stalker [1] in the UK and Lawrence and Lorsch [2] in the USA.

This research has led to a clearer understanding of the factors to be taken into account. As so often in the case of management, the identification of the elements of a problem leads to no clear resolution of it due to the need to satisfy conflicting requirements. Thus the organizational structure appropriate at one point in time may be inappropriate at another time when conditions have changed. Different managers attribute greater importance to some factors than others. It is, therefore, not surprising that two divisions of a company with apparently similar problems may have different organizations, or, as happened recently in one major company, a matrix organization was being adopted by one division at the same time as it was being abandoned by another.

This chapter will examine the main factors having a bearing upon the innovation process, look at some of the organizational structures fashioned to meet these needs, and finally evaluate how likely they are to be effective in differing situations. We shall conclude that a variety of organizational structures have merit, but each of these structures is likely to be more appropriate in some situations than others. Furthermore, it should not be forgotten that the structure by itself does not make an effective organization. The best theoretical organization will fail if unsupported by effective managers, and conversely, good management can obtain results even when working within an unfavourable organizational structure.

Definition of an organization

What do we mean by 'an organization?' In some minds, the word is equated with the company organization chart or 'family tree', which is supposed to represent a model of how the human hierarchy within the firm operates, who is responsible for making what decisions, and the superior – subordinate relationships. Frequently, this is referred to as the 'formal organization' to distinguish it from the 'informal organization' of relationships between individual managers, often cutting across the formal demarcation lines necessary 'to get the job done'. Thus, for an examination of organizations to be meaningful, it must look behind the formal pattern to the actual flows of information which determine the progress of the business. Any barriers erected to prevent these information channels which carry the life-blood of the business, information, through the body of the firm must be removed. Some system of formal relationships is, of course, essential but to be effective it must meet, realistically, the needs imposed by the tasks confronting the managers, always complex, but rarely more so than in R & D.

Lippitt [3], representative of the behavioural scientists' approach, defines an organization as, 'the arrangement of people in patterns of working relationships so that their energies may be related more effectively to the large job. The need for organization arises from the problem of dividing labour and decision making in relation to large tasks, and the need for coordinating both with respect to available energies and resources.' He proceeds to identify four systems within the organization – authority, as the division of labour, with an emphasis on the processes of work flow; personal likes and dislikes, which determine the informal organization; communication as the criteria of who talks to whom, for what purpose and with what effect; and power and control, based upon such factors as rewards, coercion, expertise and personality. These characteristics of any organization are clearly recognizable in the R & D setting. But Lippitt is typical of the organizational theorists in placing his main emphasis upon the accomplishment of a well defined organizational task. However, in the early stages of the innovation process, the tasks are far from clearly defined and consequently much of the organizational theory loses relevance. Later, as development of a project progresses, the objective does become increasingly more easy to define clearly as major uncertainties are resolved and priority can then be given to the efficient translation of project investment into earnings.

Before the development stage is reached, the idea for the new product or process must be conceived and its feasibility established. The major resource employed for this is human, and success depends upon the technological competence and skills of the R & D workers, their creativity, their motivation and identification with the project, all attributes which require a different organizational setting from development work. The ill-structured and poorly defined nature of many of the problems imposes great demands upon managerial skills.

Behavioural scientists find that the more programmed tasks are, the greater should be the organization's concern with structure. Conversely, the less the task can be programmed, the greater must be the concern with people. Thus the research end of the R & D spectrum places greater demands upon the manager's social and interpersonal skills than development where greater reliance can be placed upon formal structures. It is important for R & D managers to appreciate these differences and to develop the appropriate structure and leadership style. This problem is accentuated by what some writers have referred to as the 'thing' rather than 'people' orientation of the average technologist.

We shall start therefore, by examining some of the personal and career problems facing R & D management, since these are essential inputs in the design of appropriate organizational structures.

The human resource

The values and orientation of R & D staff

Qualified research workers usually enter an R & D department directly from university where their studies have been concentrated in one subject discipline. In the case of a PhD, this specialization may have lasted for up to seven years. Consequently, the young research worker will tend to identify with his discipline. If asked what he does for a living, he will describe himself as a physicist or a biochemist, whereas his contemporaries elsewhere in the business are more likely to mention their function and probably name the company as well. He is likely to relate his career goals to advancement in his own discipline. We have seen earlier that even senior managers in R & D may not identify closely with their company's business objectives and do not view their personal careers in terms of advancement within a particular company. It is worth noting, however, that in spite of the views they state, there is little evidence to show that the career mobility of R & D managers is any greater than for other professionally qualified people. What is more important, is that their decision-making may reflect a non-commercial orientation, albeit subconsciously.

Rejection of the business culture and the profit motive is clearly undesirable in the employee of an organization for which profits is the major objective. It does not follow, however, that a professional orientation is undesirable and should be discouraged. The company's long-term technological health depends upon the existence of a group of people developing within their own disciplines and whose interests and motivations ensure they keep abreast of the latest developments in science and technology. The transfer of technology from the outside world of science depends upon contacts with universities and other professionals in their own disciplines.

Although long-term survival may depend upon the acquisition of new knowledge for possible future applications, the short-term economic health of the company depends more upon the purposeful application of today's knowledge in a practical and commercial form. Striking the correct balance between these two must inevitably be largely a matter for individual judgment. Nevertheless, company strategy should indicate in general terms where that balance ought to lie; managerial attitudes must be developed and the organization structure designed to further that strategy.

Non-technological managers experience difficulty in understanding the professional orientation of the research worker. They may attempt to change it, although if successful, the change may not always be in the company's long-term interests. Such attempts are, however, likely to fail. Barnes [4] reporting on studies carried out by the Division of Research at the Harvard Business School concludes: 'Very briefly the findings question the usefulness of management orientations which stress profits, productivity, and practicality to the exclusion of other values. Organizational values, according to the findings of this study, may be over-stressed and self-defeating in technical groups.'

Thus we see that the organization must:

1. Enable the development of professional expertise within its staff and give recognition for the value this has in technology transfer, the technological development of the individual and the acquisition of the knowledge base for potential projects in the long term.
2. Identify and use technological gatekeepers who appear to play an important role in technology transfer.
3. Develop a commitment to the short-term application of technology through projects which meet the economic needs and strategy of the company without alienating the R & D staff by undue emphasis on profit and organizational objectives.

Transition from research technologist to research manager

As the research worker grows older he reaches a stage when he faces the difficult problem of deciding the direction his future career should take. Some will resolve it by moving out of R & D into another function within the same or another company. A decision to stay in R & D means a decision either to remain primarily a technologist for the rest of his working life, or to devote an increasing proportion of his time to managerial duties. This is a vital personal issue, but it is not one which the individual can resolve by himself, since his circumstances are largely dictated by the situation and organizational needs.

Often the research worker finds that his time is gradually becoming occupied by managerial activities without a conscious decision being taken to abandon the path of pure technology. This is accompanied by an appreciation that his technological knowledge is becoming obsolescent, partly because younger people emerging from the universities are more conversant with the latest scientific advances and partly because the time demands of his managerial role prevent him from keeping abreast of the latest developments.

As a manager he is likely to be responsible for work in a variety of disciplines in most of which, if not all, his subordinates will have a deeper knowledge than he has. His previous experience will slowly become less relevant to the present situation. All managers must come to terms with this, but few find it so difficult as the technologist whose training and value systems are based upon respect for learning and the acquisition of knowledge. Furthermore, he is unlikely to have received any preparation or training for his managerial role for which he may indeed have neither inclination nor aptitude. Thus frustration sets in.

Alternatively he may resist the change, and opt to continue a technological career. This solution also presents him with difficulties for status in an industrial culture is associated with a managerial role. While his professional attributes are recognized by his colleagues in the work situation this is less likely to be so in his relationship with outside social circles and sometimes also with his family.

He must also accept that much of his knowledge is becoming out of date. In spite of increasing specialization the sheer volume of new information can lead to a rapid obsolescence of an individual's knowledge and ignorance of work being done in other organizations; this can give rise to a waste of scarce resources through the duplication of effort carried out elsewhere and already reported in the literature. Edmund Leach puts it this way: '. . . it is, by and large, the man under 40 who "knows what is worth knowing" − the computer man, the microbiologists, the ethologists, the radio astronomers: in such fields anyone with a white hair in his head is hopelessly out of date' (Reith Lecture 1967). This is even more true in the 1990s when technological obsolescence may set in before the age of 30.

The organizational implications of the position of the senior R & D workers are thus:

1. Acceptance that the organizational association of status with a managerial role is not wholly consistent with the internal values system of the R & D department where professional excellence may command greater respect.

   Some companies attempt to reduce this dilemma by means of a dual hierarchy providing promotion prospects on a 'technical ladder' supposedly of equal status to the parallel 'management ladder'. These attempts have not always been an unqualified success. Shepard [5] noted the following problems from observations in a number of companies where a dual hierarchy had been adopted:
   (a) Difficulty in defining the role of positions on the technical ladder.

  (b) The use of the technical ladder as a reward rather than an opportunity.
  (c) The use of the technical ladder as a shelf for senior staff found lacking either scientifically or managerially. A position on the technical ladder was consequently seen as a proof of inadequacy.
  (d) Ambiguity of the technical ladder as a status symbol.
  (e) Removal of the scientist from the main stream of activity encouraging him to leave the company.

In spite of the problems dual ladders are found in many organizations. Gunz [6] in a survey of 33 UK companies found that 60% had a dual ladder but concluded that adopting it for more than a small minority is likely to defeat its aims.

It will be seen later that the matrix organization which has grown in popularity in recent years can achieve many of the same objectives as a dual hierarchy.

2. Technological staff with the right attributes for managing within the R & D environment must be formally selected and trained for their new role. This implies that the characteristics of the good R & D manager are different from those of other managers. It does not mean that technological competence is a substitute for managerial ability. Failure to appreciate this simple fact contributes to the low standards of management found in some laboratories.

   The appointment of a non-technical administrator is not the answer. If the manager is to be effective he must command the professional respect of his subordinates. A previous career of outstanding scientific achievement is not necessary for this, but the manager must possess sufficient of an R & D background to demonstrate his understanding of the professional problems of his staff.

   Formal development of managerial knowledge, skills and attitudes for the R & D manager is particularly important since his previous experience will have exposed him less than other managers to business values and management thinking. Until recently, only a few major UK companies included R & D managers in their management training programme, partly because the need has not been widely recognized and partly because the existing educational programmes did not appear suitable.

3. Career opportunities should be made available outside the R & D department. The largest number of highly qualified staff in the company are found in this department, yet few will be scientifically outstanding and the managerial opportunities within the R & D department are usually limited. Their career prospects can be extended by offering alternative employment where their intellectual powers can be brought to bear on problems in other business areas where trained minds of their calibre are rare.

   Transfer of some staff to other departments also assists the internal R & D problem of organizational renewal since the posts they vacate can be filled by younger people with the latest knowledge and nearer the peak period of their creativity. This is a practice widely adopted in high-technology Japanese firms and contributes to the rapid diffusion of technological change throughout the organization at all levels.

The role of the research director

The research director must bridge the interface between the technical interests and orientation of his own staff and the business orientation of top management. This is one of the most demanding roles for any member of the senior management team. He must at all times:

1. Be the operational head of his department and give technical leadership to his own staff.
2. Interpret the objectives and strategies of the business for his staff and develop their understanding of them so that their decisions will be consistent with company policy. (In this he must be careful not to give the impression that he identifies too closely with top management's attitudes and values if he is to retain their unreserved cooperation.)
3. Represent the R & D viewpoint to top management and ensure that it is reflected in corporate objectives, strategy and policies.
4. Play a full part as a member of top management as distinct from his specialist role as their adviser on technical matters.

The role of the research director described in the above terms places a considerable stress upon the individual who must identify with two quite different cultures and also interpret each to the other. He can be likened to a 'business gatekeeper' through whom the main channel of information between R & D and top management flows. Whereas there are likely to be a number of 'technological gatekeepers', the research director must perform his gatekeeping function largely unaided. Equally important is his educational task. First, he must develop in his staff an understanding of the objectives of the business and a willingness to reflect them in their decision-making. Secondly, he must create in the other members of the top management team an appreciation of the process of technological innovation and the contribution it can make to the company's future. We have seen already that top management's support is vital for successful technological innovation. The research director plays an important part in forming these attitudes or in changing them.

If one examines the history of successful technological companies, it is found that early growth was normally based on the contribution of one gifted man, frequently a technologist though not necessarily highly qualified. This creativity or energy provided the product which sustained the company through its early years. At that time the technology shaped the business and little attention was, or needed to be, paid to wider strategic considerations. This remains largely true where technological leadership and an offensive strategy remain the cornerstone of a company's business philosophy. In these circumstances, the natural technological orientation of the research director closely matches the requirements of the business as a whole.

The number of companies which aspire to technological leadership as they reach maturity is, however, relatively small. Strategies become more defensive and few of their innovations lead to new products or processes departing substantially from current practice. The relative importance of marketing, production and finance in company policy grows and it is no longer appropriate for the research director to remain solely the 'chief technologist'. This situation, the norm rather than the exception, calls for the full exercise of his role as a member of the top management team. Few men can measure up to the demands of all aspects of this job. It is important, however, that the research director recognizes the need, appreciates his own deficiencies, and delegates to his staff certain tasks where his own abilities are weak. Often it is the corporate role which he would wish to relieve himself of, but this is the one aspect of his job which he alone can discharge.

Leadership style

Different managers have different styles of leadership. Some are able to adjust their style to the needs of changing circumstances, but for most it is a reflection of their personality and values which are unchangeable. Most sociologists would agree that there is no 'best' style of leadership, but that the most appropriate is a function of the social environment and the nature of the tasks to be performed. The changing educational and social environment of business probably explains the increasing support for participative management methods, although their adoption in industry has not always met with the degree of success hoped for.

In general, a participative style does seem most appropriate for R & D, although the 'research' and the 'development' environments are sufficiently different for it to be wrong to assume that what is right for one is necessarily right for the other. If indeed, the conclusion is that 'research' and 'development' do call for different styles in a particular industrial setting, management must then decide whether it is desirable for an individual project to remain the responsibility of one manager throughout its duration.

The research manager's contribution to his projects comes more from his managerial than his professional skills. His decisions involve judgments over a range of disciplines in which he cannot expect to possess expert knowledge. Thus, compared with most other managers he is much more reliant upon support from his subordinates. Judgment, always important in management, is never more so than in research when the uncertainties are so much greater. This situation favours a style which is truly participative, but it does not imply abdication. Summarizing research into the effectiveness of three styles of leadership – *laissez-faire*, participative and directive – in a number of laboratories, Baumgartel [7] concluded that a participative style ranked significantly higher than others in respect of overall satisfaction with the leadership, attitude towards the laboratory director, the importance attached to a research orientation, and the extent to which the job provided for a research orientation. This conclusion is consistent with the earlier statement that unstructured problems call for a 'people' management orientation.

By contrast, the more highly programmed nature of development which places greater emphasis on structure than people enables management to become more directive. It is an open question whether a directive style is desirable in development, but there is little doubt that it is entirely inappropriate in the research laboratory.

The work to be done

Apart from the very small amount of basic research undertaken in the largest commercial laboratories, the formally planned R & D effort is devoted to work on reasonably well defined tasks or projects with specific objectives. The individual project can be considered in three stages, each of which may have different organizational implications:

1. Conception.
2. Justification.
3. Implementation.

Conception – creativity

A prime organizational objective must be the positive encouragement of creativity through the recruitment, retention and development of creative people without whom there will be no innovation. However, we have observed that many organizations unwittingly achieve the opposite effect through an over-insistence on formal procedures and detailed justification for ideas still in their formative stages.

Too little is known of the process of creativity for it to be possible to prescribe an organizational climate for its encouragement. Nevertheless, it is possible to identify inhibiting features in the working environment which should be removed.

Justification – selection and evaluation procedures

Examination of the requirements for project selection and evaluation stressed that the processes involved were as important, if not more so, than the techniques used. The culmination of any project is a product or process which serves the business by furthering its *strategy* through satisfying a *market*, in a form which can be *produced* at a *profit*. Thus at every stage, including selection, the project is not of concern solely to the R & D department, but also top management, and the marketing, production and finance functions.

Organizational structure, which is 'the arrangement of people in patterns of working relationship so that their energies may be related more effectively to the large job' (Lippitt), must therefore permit and encourage the involvement of corporate, marketing and financial staff in the selection and all subsequent stages of a project. Thus, consideration of an organization for innovation cannot be confined to the R & D department in isolation, since the process extends throughout the company. Whatever form of organizational structure is adopted some arrangements for the effective association of the other functions must be devised. The structures we find in practice range from the complete restructuring of the company organization to meet the needs of technological innovation at one extreme, to a strictly departmental organization linked loosely to the other functions.

Implementation – feasibility to end-product

Management is the art of making things happen. The agent for bringing this about is the manager utilizing resources entrusted to him for the achievement of an objective, which in the case of R & D is the successful completion of a project on time at minimum cost. Because he is achieving something new, progress of the project depends upon his judgment and his decisions. No one else has the same intimate knowledge of the project. If this knowledge is allied with sound judgment, good decisions, drive and determination, there would seem to be no benefit to be derived from restricting his total responsibility for all aspects of the project's direction. The research evidence showing the importance of the 'project champion' supports this argument.

Total delegation of project responsibility demands a degree of organizational courage few companies are prepared to take. Can one be confident that the project manager's judgment and decision-making ability are to be trusted? The risks are great, particularly when the future of a major project is at stake. To what extent can an organization afford to allow one man to control the use of costly resources when he is often a person with little regard for accepted procedures? The natural tendency is always to apply detailed controls to ensure the custodian of these resources uses them wisely, even when the logic of the situation points to the project manager as the person who should know best. This dilemma lies at the centre of much of the controversy surrounding the best organization for innovation. A few companies have gone the whole way in freeing some of their project managers from virtually all organizational constraints. In other companies there is a growing awareness of the importance of entrepreneurial qualities, and an increasing willingness to experiment with new forms of organizational structure.

Thus we see that the organizational structure has considerable effect upon the management of a project. The form adopted by a company must depend upon the relative importance attached to the conflicting demands of:

1. Freeing as far as possible the manager who identifies with the project, is most concerned to make it a success, and has the greatest knowledge of it; *and*:
2. Ensuring that the expenditure of company resources is controlled tightly.

Industrial characteristics

All companies and industries are unique. The individual characteristics of the particular environment in which innovation has to take place have an important bearing upon the purpose to be served by the organizational structure. There is little in common, for example, between the pharmaceutical research laboratory attempting to synthesize a new drug and the development of an advanced aerospace project. These differences must be appreciated, and the organization designed to meet the peculiar requirements imposed by the characteristics of its own technological environment. Ansoff [8] has identified five factors which he suggests make up the 'technological profile' of a company. All five have a number of organizational implications, a few of which are noted in the following paragraphs:

1. *The R & D mix.* We have already seen that the organizational requirements for research may be quite different from development. The difference is often clouded by the higher status associated with 'research'. Laboratories which are almost entirely devoted to development are sometimes called 'research laboratories' and organized as if indeed they were engaged in research. Thus it is important that the nature of the tasks undertaken are analysed dispassionately and the organization matched to the reality of the mix between research and development existing in the company.
2. *Coupling between departments.* The degree of coupling between departments results from the organizational structure. The interfunctional nature of most major R & D decisions has been stressed throughout this book. It has also been suggested that, in general, the amount of interaction between departments (i.e. the coupling) is less than desirable. However, the nature of the projects has a relevance. A close coupling between R & D and marketing, for example, would not normally be found or thought necessary where a detailed specification can be written for a major project with a gestation period lasting several years, e.g. nuclear reactors or supersonic airliners. Thus the organizational structure, while always allowing for some degree of coupling, should reflect the actual needs of the situation.
3. *Average product life-cycle.* A short product life-cycle indicates rapid technological advance or an unstable changing market. Product development must reflect the latest technology on the one hand, and the current market requirements on the other. This demands an organizational structure which fosters a high degree of technology transfer and interaction with marketing. In a mature industry with a long product-life cycle, these organizational objectives are less important, the time pressures are less, and it is possible for a project to move from one stage to the next on a well defined path without the same degree of close association between R & D, production and marketing, as is necessary when corners have to be cut to put a new product onto the market quickly.
4. *Research and development investment ratio.* A high investment in R & D

indicates that it is expected to play a significant part in the future development of the business. The needs of technological innovation are more likely to exert a major influence on the organization structure of the whole company, and it becomes more important that the R & D programme is closely tied into the corporate objectives.

5. *Proximity to the 'state of the art'.* When the technology is close to the frontiers of knowledge it may have a major impact on the business. This characterizes an offensive strategy as well as a high research content in the R & D mix. The organization must be flexible and able to respond rapidly to a major discovery. Thus, close top-management involvement is also necessary.

The organizational structure at a point of time is normally found to have adapted to the traditional technological profile. Rarely is one concerned with the design of an organization from scratch. But no company remains static for long. The importance of thinking analytically in terms of a technological profile is that it becomes possible to identify changes, interpret them, and adapt the organization to meet the new needs.

Coordination and communication

An organizational structure must simultaneously serve two conflicting requirements, in that it must both separate and integrate. It must delineate the role of each person in the company and distinguish between one individual's responsibilities and those of others. This is often achieved through formalized job specifications although the responsibilities defined by it must be interpreted liberally in an R & D setting. Whilst it is essential that the specified tasks are undertaken it should not be regarded as a constraint since it is impossible to define precisely what each person must do in a fluid and complex situation. The interfaces must be bridged and coordinated so that the total effort is integrated. However, much of this results from informal contact which the organizational systems should encourage rather than inhibit by imposing barriers. It is much easier to specify the physical work to be done, for example in production, than it is to specify the information flow requirements which are a major element of the innovation process. Moreover, it has to be recognized that many technologists are insular and do not communicate freely, particularly across functional boundaries. Thus the organizational system aided by technical management must be designed to stimulate the coordination of tasks and facilitate the flow of information throughout the company. The structure should be an enabling mechanism which reinforces and encourages the processes required for speedy and effective innovation. All too often it has been designed for other purposes and is used to maintain the status quo.

This coordination and communication must be achieved in two directions – vertically and horizontally. In the vertical direction it must facilitate the linkage between the strategic management of the firm at the top and the

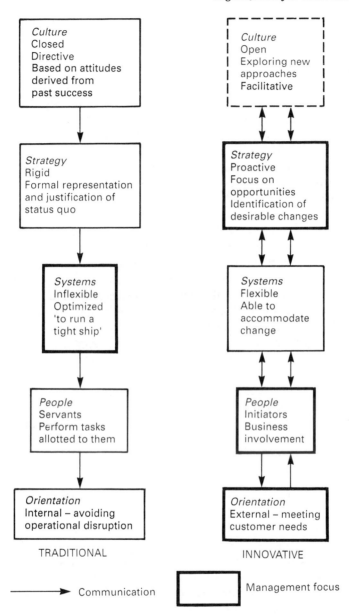

7.1 Traditional and innovative managerial emphases

technological creativity and entrepreneurship at the bottom, as discussed in Chapters 1 and 2. Figure 7.1 draws a distinction between traditional and innovative organizations in respect of their culture, strategy, systems, people and orientation. They differ considerably, and it will be noted that whereas communication in the traditional organization is limited and almost entirely in a *top-down* direction, that in an innovative organization the volume is high in both *top-down* and *bottom-up* directions.

The corporate culture, reflecting the attitudes and values of top management, is a key factor. An innovative organization is likely to result when it is based on a belief in technology, growth and the need for change. The emphasis on the role of the champion, the poor representation of technology in formulating corporate strategy, and the search for means of overcoming internal barriers which are revealed by innovation research and reported in the literature indicate that these conditions are absent in many companies. Hence there is an assumption, often well founded, that much of the management of technology consists of introducing innovation in an alien environment.

The research of Twiss and Goodridge [9] shows that most of the features of the innovative organization are found in successful technological companies, particularly in Japan. Much R & D, however, is conducted in large mature firms where the situation approximates to that typified as traditional. This makes the task of innovation much more difficult although not impossible. It must be achieved in spite of rather than with the assistance of the formal corporate systems. Consequently informal arrangements must play a much larger role and it is incumbent on the technologist to take the initiative in overcoming the shortcomings of the environment within which he is working.

Equally important is the need for good horizontal coordination and communications between the functions at similar levels in the hierarchy. This is the prime concern of those responsible for the introduction of specific innovations. Failure to establish this is a major source of development problems leading to delay and cost over-runs and is a feature of projects which regard innovation as a sequential process of R & D, design, production and marketing. This will be a major concern in the discussion of organizational structures in the following sections.

Summary of aims to be achieved through organization

The discussion so far has identified a number of considerations to be taken into account when designing an organizational structure to meet the needs of technological innovation. It is seen that it is difficult to reconcile some criteria with others having opposite oganizational implications. An organization must, consequently, be a compromise between these conflicting forces and the form it takes should depend upon management's considered judgment of where emphasis ought to be placed. The most important of the factors the organizational structure should take into account are:

1. The *long-term* development of staff and technological expertise, *vis-à-vis* the *short-term* needs to apply technology for immediate commercial ends
2. Transfer of *technological information* into the company from outside sources and transfer of *corporate policies* into R & D.

3. The association of *marketing, production and financial* staff to the full extent necessary.
4. Preservation of a high degree of *autonomy for project managers*, while maintaining *corporate control* over the resources invested in a project.
5. A *leadership style* appropriate to social and organizational needs.
6. Recognition of the company's *technological profile*.
7. The need to stimulate *creativity* and motivate staff.

Organizational structures

It is now possible to examine a number of different organizational structures and assess their merits in relation to the criteria we have identified. Several of these structures are similar in concept, differing mainly in the importance placed upon the authority of the project manager. No clear demarcation exists between project management, matrix organization or venture management, although the use of these terms implies an increasing degree of project manager autonomy, accompanied by a lower emphasis on the scientific discipline as one moves towards venture management. Intermediate forms also exist under a variety of titles – one ICI division, for example uses the term 'business manager' to describe a project manager who, while broadly working within a 'matrix organization', possesses some of the characteristics of the 'venture manager' without the same degree of independence.

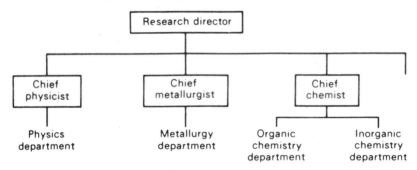

7.2 Laboratory organization by scientific discipline

The name 'project manager' also leads to confusion since it is a generic title used in most organizations for the head of a project. It is also used in a more limited sense to describe a particular form of organizational structure.

Organization by scientific discipline

Many laboratories follow the university practice of organizing their work according to scientific or technological disciplines. Thus there will be departments of physics, chemistry or metallurgy, often subdivided into further subspecializations (Fig. 7.2).

The graduate entering an industrial laboratory has little difficulty in identifying with this type of organization, which possesses features aiding his individual career development within his own discipline. The advantages from his viewpoint are:

1. Familiarity with the organizational pattern eases his absorption into the industrial enterprise.
2. Identification with colleagues in a specialist group from whom he can obtain professional assistance and intellectual stimulation.
3. Employment in an area for which he has prepared himself through his earlier studies.
4. Knowledge that his professional competence will be assessed by a superior with a similar background who is fully competent to judge his performance.

The discipline structure is well suited to the acquisition of new knowledge in a specialist field. It is the logical structure for a university, but a commercial enterprise, even when knowledge acquisition is accepted as a major objective, has the overriding aim of turning technology, from whatever source, into a form having a commercial value. Concentration on the disciplines detracts from the importance of the project, the vehicle for technological innovation. It also makes it difficult to discriminate between the useful and the interesting. Pursuit of the interesting for its own sake does, of course, occasionally lead to outstanding discoveries or a serendipitous outcome. Although it would be unrealistic to ignore this entirely, there seems little justification for basing the organization of a commercial laboratory upon such uncertain and rare occurrences.

It is also difficult to inspire any sense of urgency into an organization which mirrors that of the university where time pressures in research are relatively less important than the quality of the result. In an industrial laboratory a quick practical solution is usually preferable to a delayed best solution.

Perhaps the most serious drawback of the discipline approach is that it implies that innovation is most likely to result from pushing back the frontiers in a specific discipline. More frequently, however, innovation stems from a combination of advances in several often unrelated disciplines and technologies. Cross-fertilization of ideas, so important if this association is to occur, is much less likely when the organization fosters specialization. Creativity, also, is likely to be inhibited when staff are grouped in a way which discourages challenges to the accepted orthodoxy.

Organization by discipline does not appear well suited to the purposes of technological innovation, although it is highly suitable for the acquisition of new knowledge and the satisfaction of the professional wishes and development of R & D staff. Nevertheless, this structure is frequently observed in practice. Mansfield [10] noted that 47%* of his sample of nineteen US laboratories

---

* Individual industries showed wide variations, possibly reflecting their technological profile and the number of scientific disciplines involved in any one project, viz.: chemical, 57%; drug, 80%; electronics, 33%; petroleum, nil.

were organized by discipline; although comparable statistics for Europe are not available, there is little reason to doubt that they would show a similar result. In central research laboratories, where research forms a significant proportion of the work, the advantage of a discipline approach may well outweigh the disadvantages. Yet it must be questioned whether the extent to which it is found does not reflect more the personal inclinations and needs of the R & D staff than the needs of the organization they serve.

Project management

Work intended to culminate in specific new products or processes consists of individual projects, each with a clearly defined objective. Contributions from a number of disciplines must be coordinated. The problem of attempting to achieve the necessary day-to-day coordination through a system of committees are such that it is usual to appoint one man, the project manager, whose role is to ensure the successful completion of the project. He is likely to be a member of the discipline making the major contribution to the project, although this need not necessarily be the case.

In the simplest form of project management, he acts solely as a coordinator, although the staff from his own discipline working on the project may report to him directly. Staff in other disciplines remain responsible, both professionally and managerially, to their own discipline heads. The project manager works within a system of committees, but is free to take routine decisions affecting his project. Policy decisions affecting the project and the allocation of priorities between projects are usually reserved for a committee chaired by the research director, attended by the discipline heads and possibly also the project managers. Less important decisions are usually taken by a project committee, chaired by the project manager, with representatives of the disciplines involved.

The project manager may be expected to identify fully with the project; sometimes he will be the only person who does so, but he is restricted by having no direct managerial authority over the staff through whom he must obtain his results. The continued effort of this staff and the acquisition of the resources he requires depend to a great extent upon the project manager's political skill. Having little direct authority, much of his time is devoted to discussions with staff in other disciplines and attending committee meetings. When things go wrong, he can often blame circumstances beyond his control.

In spite of the somewhat ambiguous position of the project manager, this form of organization often works well in practice. Some years ago, the R & D laboratory of the Beecham Group successfully moved to project management from an organization by scientific discipline. They experienced no problems with the previous heads of departments, three of whom became project managers. At the same time, a scientific career ladder was introduced for good research workers without the ability to manage projects. The demands placed upon the project manager are illustrated by the experience of this company whose major difficulty initially was in finding sufficient people possessing the qualities necessary for effective project management.

Organization by product line

A company's business may naturally divide into a number of sub-businesses, each selling product lines of similar products or serving the same customers. A large company often adopts a multidivisional organization, where each division has a high degree of autonomy. R & D can be organized to match the

(a) *Centralized research and development*

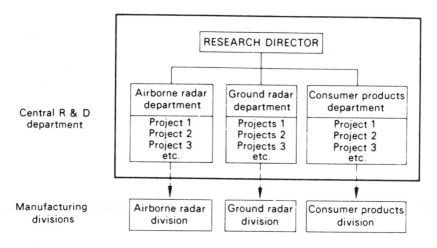

(b) *Divisionalized research and development*

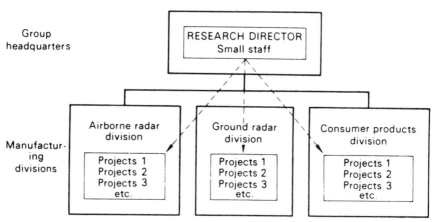

7.3 Organization by product line

divisional structure, either within a central laboratory, or by divisionalizing the R & D programme (see Fig. 7.3(a) and (b)).

The subject discipline orientation is de-emphasized in the product line organization, although the individual laboratories may still be organized by discipline, in which case there may be a degree of duplication of effort. The responsibility of the research director is likely to be delegated to a greater extent than in most other organizational forms, particularly when he has no direct authority over divisional laboratories reporting directly to their own divisional managing director.

Because of the linking of R & D with a smaller product line or group of customers, this organization strengthens the marketing orientation of the R & D laboratories.

Matrix organization

The matrix organization is a development of project management designed to avoid the disadvantages of the simple project management organization, by separating clearly the managerial and professional responsibilities for the project (Fig. 7.4).

The R & D director has reporting to him the discipline heads, who are responsible for maintaining professional standards within their own disciplines and the career development of their staff, as well as the project managers who

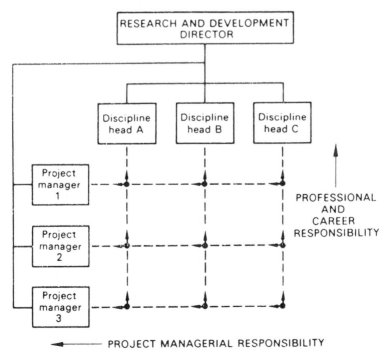

7.4 Matrix organization

are responsible for the progress of their projects. The individual research worker is responsible to the project manager for the day-to-day progress of his work on the project, but is responsible to his discipline head for the professional standard of his work and to whom he looks for career guidance and development. This organization appears to conflict with two of the rules of traditional management teaching:

1. No person should report to more than one manager.
2. Responsibility and authority must be equated.

These rules, desirable though they remain, stemmed originally from the military staff colleges dealing with problems without the complexities of modern industrial situations. Nevertheless, an organization which ignores them inevitably introduces an element of ambiguity placing a heavy demand upon the managers concerned. It can be argued that it is for the exercise of such qualities that the manager is paid. Support for this view comes from the success of many organizations based upon this principle.

Compared with simple project management, matrix organization emphasizes the managerial responsibility of the project manager. He is held accountable although he commands only task authority over all staff working on the project. If difficulties arise with the section heads, he must resolve them. Although he has direct access to the R & D director, frequent recourse to this channel to resolve interpersonal problems must be kept to a minimum if the willing cooperation of the section heads is not to be sacrificed. He alone can ensure their continued willing cooperation.

The merits of the matrix organization will now be examined in relation to the achievement of company objectives, the project manageer, the discipline head and the individual R & D worker.

Company objectives

The emphasis which this organization places upon both managerial and professional needs is a compromise ensuring the energetic pursuit of project objectives while safeguarding the professional interests of the majority of the staff and the retention and extension of the company's long-term technological capital. It is a delicate balance, and success depends upon the existence and development of the required managerial skills and attitudes.

The principles of the matrix organization can be extended to embrace other functions as well as the disciplines represented in R & D. Figure 7.5 shows how the involvement of engineering, production, marketing, and finance staff necessary for the full innovation process can be embodied within a matrix organization. As the project progresses, so the departmental representation in the project team will vary. In these circumstances, smooth transition between departments may be achieved by changing the project leadership first to a production and later to a marketing manager. Ideally, however, one is looking for a business orientation (see 'venture management') in a leader who can carry

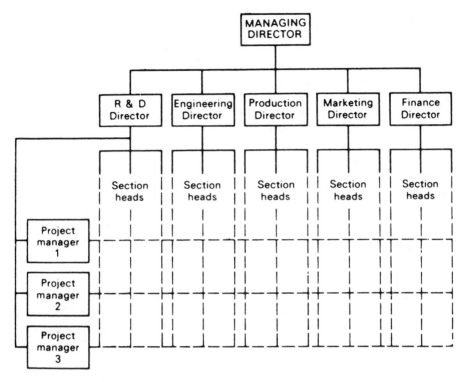

Note: 1. The matrix organization usually applies to only a small proportion of the staff in the production, marketing and finance departments. It exists for new projects and does not replace the existing organization.

2. The project managers are shown here reporting to the R & D Director. At a later stage in the development, this responsibility may be transferred to another functional director.

7.5 Matrix organization for the innovation process

through the project from start to finish without the discontinuities inevitable when the manager is changed.

The project manager
The project manager's attention must be focused upon the management and progress of the project, rather than on the personal solution of technological problems. He is an integrator, who interprets information from technologists in several disciplines and managers in other functions. He is a decision-maker, exercising his judgment over the total project. He is a motivator, who gains the cooperation of his staff in implementing his decisions. Appointment to project management within a matrix organization marks a definite career step from technology into management which is most probably irreversible.

The project manager's close identification with his project makes it difficult for him to assess it objectively. He becomes the 'project champion'; any criticism of the project he interprets personally. However, this personal commitment, which is of such benefit to the project when it is progressing satisfactorily, can have the opposite effect when new information indicates that the project should be terminated. The project manager usually resists termination with all his determination, not only because of his belief in the project, but also because he may feel insecure, regarding the termination as a personal failure. Unless the status and management career pattern for the project manager is made explicit he may consider he has good cause to feel insecure. In this case, he is likely to be torn between his project involvement and a desire to maintain a foothold in his own discipline to which he can retire if his project management proves unrewarding. This can be more significant when he is one of a number of project managers working on relatively small short-duration projects. The termination of a major long-duration project may mark the end of a phase in his career; his next appointment must not only be planned but also communicated to him. These considerations do, of course, apply to all project managers, whether or not they are working within a matrix organization.

The discipline heads

The discipline heads often present the greatest problems in a matrix organization structure, since they may feel divorced from 'the action' and isolated from the project enthusiasms of their staff. Compared with a previous organization, the discipline head may interpret the change in his position as a loss of power and status.

The justification for a matrix organization is the importance placed by the company upon both managerial and professional criteria. Discipline heads are relieved of many of their managerial responsibilities and freed to devote more of their time to developing the future contribution from their subject area. This goes some way towards satisfying the career needs of the senior scientist or technologist who has no desire or ability to move more towards management. However, the fears of the discipline heads will only be allayed if their equality of status with the project manager, implicit in the organization structure, is also made explicit in the behaviour of top management towards them. Often this is overlooked in the emphasis top management place on the achievement of the short term and hence the project manager.

The individual R & D worker

The individual R & D worker finds himself within a small closely knit team working towards a specific and tangible objective. Being the expert in his discipline he is awarded a higher status within the multidisciplinary project team than if he had been working with a number of other scientists within his own specialist field. Yet he retains his links with his own discipline, and, in particular, with his discipline head to whom he can refer professional problems.

Since the majority of R & D staff like working to specific objectives, the matrix organization has been well accepted by R & D workers.

Organization for transfer from R & D to production

The matrix organization was seen to permit the gradual transition of a project from R & D to production without the major discontinuity involved by a formal transfer between departments. This is achieved by changes in the composition and perhaps also the leadership of the project group. In the other organizational structures described so far there is no provision for bridging this gap between R & D and production which can constitute a major cause of programme dislocation. Frequently, early production models exhibit faults not experienced with the prototypes, and production plants suffer from problems unknown in the pilot plants. Speedy resolution of these problems is hampered by poor communications between the two departments and mutual recrimination. This is essentially a communication problem associated with the organizational structure whereby production staff are denied access to much of the R & D data and skills never committed to paper. Similarly, R & D are often unaware of the problems of reducing the outcome of their projects to a producible and marketable product.

Several organizational devices have been adopted to overcome this problem – liaison groups, cross transfer of personnel, task groups, and new product departments.

Liaison groups
A liaison group consists of personnel appointed to understand the project as it emerges from R & D and resolve problems arising from its preparation for

(a) *With liaison group*

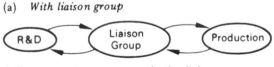

Indirect contact – two communication links

(b) *Without liaison group*

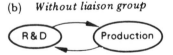

Direct contact – one communication link

7.6 Communication links

production. Since they are supposed to have no departmental allegiance, they are expected to view these problems objectively. Some companies have found that liaison groups have proved valuable where the existing relationships between departments were poor. Nevertheless, they are open to the following criticisms:

1. The appointment of a liaison group involves the employment of additional staff and leads to a certain amount of duplication of effort.

2. The communication links, and thus the possibilities of error, are doubled (Fig. 7.6).
3. The two departments are likely to become more isolationist. Burns [11] noted this tendency in his study of companies where liaison groups were also observed to be accompanied by a tendency to dissociate from the commercial purposes of the firm.

Transfer of personnel

A transfer of personnel between the two departments is another device used by a number of companies. During the final R & D stages of a project, production personnel are transferred into the R & D department to familiarize themselves with the techniques and unresolved solutions to problems experienced there. They may be accompanied by a few R & D personnel transferred temporarily to the production department when the project moves into production.

Task groups

A task group is a form of temporary interfunctional project team formed to bridge the gaps between R & D, production and marketing. The members of the task group normally devote only a proportion of their time to the project, the remainder being occupied with duties within their functions. The task group has greater direct responsibility for the project than a liaison group and bears many similarities to a project group within a matrix organization. It is, however, superimposed on the existing organization as distinct from a restructuring of it.

New development department

A new development department is a development from the task group, with a larger and more permanent membership.

Venture management

The term 'venture management' is used to describe an organization for innovation, designed to reproduce within the large company many of the attributes of the small entrepreneurial enterprise. The basic philosophy is to give maximum responsibility for the progress of an innovation to one man, the 'venture manager', who is free to use his resources with a minimum of outside interference, once the project has been given formal budgetary approval. Usually, 'venture management' is reserved for a few projects of exceptional promise and is operated alongside the existing organization.

The key to success lies with the venture manager who is appointed not only for his technological ability, but more importantly, for his attributes as a manager and a businessman. The degree of autonomy he enjoys varies with the nature of the project, but generally he has complete control of the spending of his agreed budget, including recruitment of staff and changing the deployment of his resources as he thinks fit. He differs from the project manager in a matrix organization in his increased autonomy and his role which extends beyond that of manager to businessman. His relatively small organization and short lines of communication permit considerable flexibility in planning as the project develops, for he is in effect the chief executive of a small business, having total responsibility for the R & D programme, setting up manufacture, and introduction to the market.

Venture management concentrates wholly upon the rapid exploitation of a specific innovation, thus the staff of the venture have little, if any, time or organization link with their disciplines. This is no problem for a large firm, where long-term research can be catered for elsewhere in the organization.

In spite of the undoubted success achieved by venture management, it should be used selectively for there is a limit to the number of semi-autonomous operations an organization can accommodate, as well as the supply of managers with the requisite exceptional ability.

Unlike a matrix organization, there is no departure from the rules of 'one man, one job' and 'no responsibility without authority'. In achieving this, the company is breaking up the organization into small businesses with a consequent loss of some, but not all, of the advantages derived from being part of a large company. It can also mean sacrificing what may be important career considerations of some of its staff in return for the simplicity inherent in the small organization. However, if we believe that in the past long-term scientific R & D has all too frequently been given precedence over commercial interests, we might expect more companies to move towards some form of venture management as a response to increasing competitive pressures.

Joint ventures

In recent years the number of joint ventures has been growing rapidly, a trend which is likely to continue. There are a number of reasons for this.

For some technologies the *cost of development* is beyond the financial means of any one company, hence a joint development programme to share the cost. A more limited form of cooperation is an exchange of information. Thus we find that Rolls-Royce is jointly developing new aero-engines with Pratt & Whitney, and an international group is manufacturing French and American designs under licence and has an information exchange agreement with GE.

Some developments from a laboratory may not *fit the current strategy*. Nevertheless, the company wishes to keep its options open and will provide a proportion of the risk capital to set up an entrepreneur from their own company in a new business in exchange for a stake in the equity.

Often future growth depends upon *technological synergy*. Most industries now reaching maturity have based their past growth on one or a family of technologies. These bring diminishing returns from further investment but there is scope for renewed growth from the combination of previously disparate technologies. It is, however, extremely difficult to grow a new technological capablility within an existing business where there is no expertise. Thus we are seeing the emergence of a variety of new forms of joint venture:

1. Joint ownership of new companies to exploit a technology where both parties have a different contribution to make.
2. Takeover or merger where technologies converge.
3. Pre-competitive research cooperation where two or more companies, usually in different industries, invest in joint R & D in a new technology of common interest, e.g. a chemical, a food and a pharmaceutical company undertaking joint work on the application of information technology.

Streamed management

In a large company many projects will be worked on simultaneously. This necessitates the adoption of a formal organization to ensure the orderly allocation of resources and effective project control. Inevitably this leads to some loss of urgency. Senior management, which is concerned primarily with

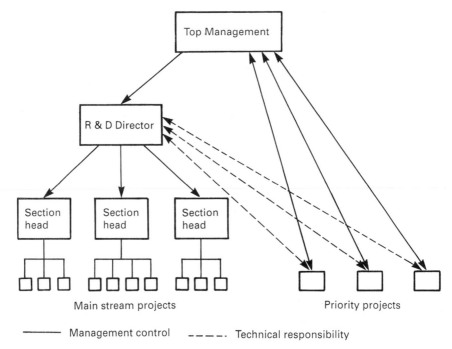

7.7 Streamed management

policy and portfolio issues, has little or no time to devote to the problems of individual projects. On the other hand, an organization based upon a large number of semi-autonomous ventures would be impossible to coordinate. Within this portfolio there will be a few projects which demand more senior management attention because of:

1. Their long-term strategic importance.
2. The need to complete them expeditiously to meet an urgent need.

One way of overcoming this problem is to categorize the portfolio into two streams. The main stream, comprising the majority of the projects, is managed within the normal organizational systems. The priority stream is removed and placed under the direct control of the R & D director or top management (Fig. 7.7). Hitachi [12] which calls this system 'Tokken', applies it to approximately 20% of the company's R & D (about 45 projects) which are monitored closely by top management. Commodore adopted this procedure for the design of their CDTV which combines TV graphics, text, animation and expanded audio capabilities with the CD player. The project team of 26 reported directly to the company president. It launched the project in February 1990 and the final product reached the shops in the remarkably short time of nine months.

Laboratory location

Many large companies have centralized their long-term research in central research laboratories separated geographically from the manufacturing divisions. The archetype is the large country house surrounded by a deer park. Here no divisional operating problems requiring urgent solution interrupt the research worker's preoccupation with the longer term. Such matter can be left to the smaller development laboratories attached to the manufacturing divisions.

The relative isolation and centralization of research has the following advantages:

1. Freedom to pursue long-term objectives without interruptions.
2. Avoidance of duplication of work relevant to more than one division.
3. Centralization of experience or infrequently used equipment.
4. Ability to build effective research groups with a wide range of disciplines.
5. Resources to investigate new technologies which do not fit into the existing divisional structure.

Against these can be set the following disadvantages:

1. Remoteness from market forces and the needs and experiences of the operating division.
2. Reduced profit consciousness.
3. Demarcation and communication difficulties between the central and divisional laboratories.
4. A tendency for R & D for the medium term to be neglected by falling between the central and divisional laboratories.

One of the dangers of a single laboratory is the likely channelling of all effort into one approach to a problem, where the uncertainties are such that alternative routes should be followed. Glaxo has recognized this danger and has organized its research into two departments, one searching for pharmaceutically active compounds which are then screened pharmacologically, the other specialized in pharmacological testing relative to important disorders and the searching for compounds to cure them.

The arguments for associating short-term R & D and product improvement with the production unit are strong. So is the case for locating long-term research near the source of knowledge, for example a university. In both situations the rationale is the same, namely the improvement of communication that comes with geographical proximity. This can lead to the location of R & D in a number of widely dispersed sites as both knowledge generation and production become more international. Howells [13] quotes the case of Glaxo where the number of laboratories between 1968 and 1990 rose from three (one in UK, two overseas) to nine (three in UK, six overseas). Shell has 15 laboratories in eight countries and the number of its central research laboratories has increased from four to eight in recent years. This dispersion, however, brings its own problems:

1. R & D becomes more difficult to coordinate.
2. Work may be duplicated.
3. Poor communication of R & D results between laboratories.
4. The divorce of long-term from short-term work.
5. Low acceptance of long-term research by operational divisions.
6. Specialist expertise in one laboratory not being available to solve the problems in another.

Some ways in which the budgeting of R & D can be used to overcome these difficulties were discussed in Chapter 2. Another solution for some of these problems is a form of technological matrix. Siemens has adopted this approach where the horizontal axes between laboratories are technical expertise or scientific discipline. The discipline head is located in that laboratory where there is the greatest concentration of expertise but he also has responsibilities for work undertaken in his discipline throughout the company. This also enables company-wide resources to be deployed to solve technical problems arising in one location. A similar matrix organization was adopted by Daimler-Benz in 1991. Central applied research and technology was decentralized into four research institutes, one for each of its operating divisions — Mercedes-Benz (cars and trucks), AEG (electrical and electronics), DASA (aerospace) and DEBIS (financial and information services). The director of each institute is additionally responsible for a 'global corporate task' covering medium- and long-term research spanning the whole company in one of the areas: IT, environment and energy, materials and production, and technology assessment.

The economic pressures of recent years have forced many companies to re-examine the contribution being made by their central laboratories. This has led to some closures and many reduced budgets, a typical reaction to short-term economic difficulties. Nevertheless, there is sound logic in separating long- and short-term R & D, provided the organization can ensure the integration of the central laboratory with the commercial needs of the company while preserving the necessary freedom to pursue long-term objectives.

The long-term nature of centralized R & D lends itself to organization by scientific disciplines, although a product line (or manufacturing division) organization is not uncommon.

Organization structure for the future

The traditional search for one type of organizational structure to meet all the needs of technological innovation has led companies to change their organizations from time to time in an attempt to find the structure best suited to their needs. We have seen that all the organizational structures examined have shortcomings, though some are better suited to some situations than others.

Better understanding of the innovation process and organizational research are leading to a realization that greater flexibility and finer tuning of the organizational structure to particular circumstances are necessary. New organizational forms may emerge. It is unlikely, however, that a solution can be devised which appears simple on paper, yet accommodates the complexities introduced by the ever-widening knowlege base, the uncertainties of the innovation process, and the randomness in the initiation of many of the most successful ideas. Management too, must learn to tolerate and adapt to the ambiguities where different criteria indicate conflicting solutions.

Examination of new organizational structures emerging in recent years shows two distinct trends, both aimed at increasing the commercial and project orientation within the large technology-based company. In one group, typified by the matrix organization, ambiguity is accepted as unavoidable and it is left to the project leader to effect a compromise and obtain the cooperation of discipline heads in support of the project. The other group aims to reproduce the simplicity of the small enterprise by dividing the larger unit into ventures or projects characterized by entrepreneurship, rapid and decisive decision-making, short communication lines, and the dedication to the project found in the smaller social unit.

The magnitude of the organizational dilemma grows with the size of the company. Remoteness is not such a problem with the small firm which is either working close to the frontier of knowledge in a highly specialized field or is orientated towards development of existing products where the professional concerns of the staff are less important. In contrast, the large companies now employ up to several thousand R & D staff, a high proportion of whom are graduates, many working in a large laboratory isolated from commercial pressures. Size, however, does provide the opportunity to adopt a variety of organizational structures in different parts of the company.

One consequence of this is the likely emergence of hybrid or multistructure organizations in the large firm. Figure 7.8 shows one such possibility, where five different types of organization can be identified:

1. *Central research* to provide the long-term technological capital is organized by *discipline* under specialist heads of department (or section).

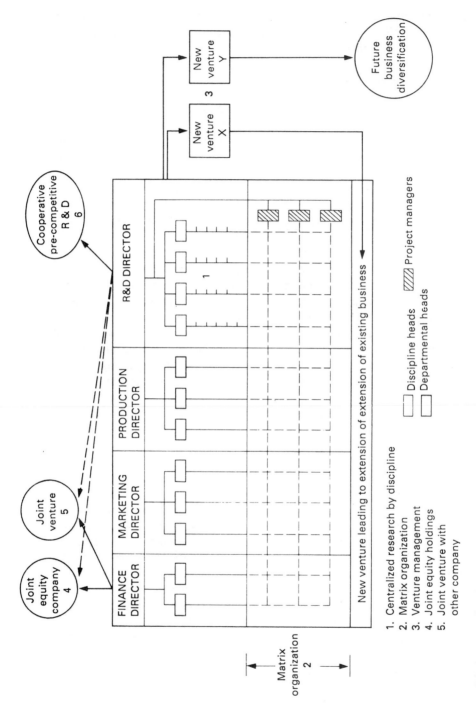

7.8 A suggested multi-structure organization for the future

1. Centralized research by discipline
2. Matrix organization
3. Venture management
4. Joint equity holdings
5. Joint venture with other company

2. *A matrix organization* for the majority of projects which leads to extensions or maintenance of the existing business.
3. *Venture management* for a few major projects of high potential. The products or processes resulting from the ventures are absorbed into the existing business or occasionally lead to a new business diversification.
4. *Joint equity* holding in companies formed to develop proposals emerging from the R & D department which, while showing promise, do not fit the parent company's strategic or market objectives.
5. *Joint ventures* with other companies, either competitors or those with synergistic technologies.

This hybrid structure largely avoids the imposition upon any part of an organization of a standardized compromise structure which inhibits the process of innovation. It accommodates different managerial styles and permits a wide range of projects with varying degrees of risk to be undertaken.

The project team

The organization only provides the structure within which a group of people can work to achieve the project objective. In order to succeed there need be:

1. Enough manpower to complete the necessary tasks in the specified time.
2. The appropriate spread of expertise either within the team or available to it.
3. People of the right calibre in terms of competence, age and experience.

Specifying the requirements as above is, in theory, a relatively simple problem in resource allocation although in practice it may be difficult to satisfy since what is required may not be available. The project manager will normally have to cope the best he can with what he is given.

A collection of talent does not in itself constitute an effective team. Paper qualifications are a necessary but not a sufficient basis for success. There are, for example, different roles to be performed in a team. Belbin [14] has categorized these roles under eight headings: chairman, shaper, plant, monitor-evaluator, company worker, team worker, resource investigator and completer. For example, the 'plant' is independent, highly intelligent and imaginative whereas the 'completer' is meticulous and concerned with detail. Team members are not chosen in relation to their aptitudes in these roles. Nevertheless, the roles must be performed and the R & D manager must assess the aptitudes of his team members and allocate tasks accordingly. Having done this he must still perform the critical managerial job of welding them into a team and motivating them to succeed.

The job of the project manager is not, however, confined to team leadership which is only part, albeit an important one, of the role. A useful categorization used by Barczak [15] considers four main roles: communication – within the team; climate setter – creating the environment within the team, including the allocation of work and the resolution of conflict; planner − setting the goals, strategy and detailed plans; and interfacer − liaison, coordination and communication external to the team.

Success does, however, seem to be related strongly to the degree of responsibility and authority invested in the project manager. For example, in a study of 540 development projects in the USA and Canada [16] there was strong evidence that the least effective were those based on a functional organization or a matrix where the project manager acted solely as a coordinator across functional boundaries.

Research by Katz and Allen [17] and others suggests that project team performance varies with the length of time they have been formed. Performance rises with mean tenure of members and is at its peak between two and four years and then falls. He attributes this fall to the development of the 'Not Invented Here' effect which inhibits the acceptance of new ideas from outside. This has implications for lengthy projects where it may be necessary to plan for change in the team membership.

Summary

The organizational objectives of technological innovation were shown to be different in character from those of other parts of a company. Objectives and tasks cannot be clearly defined in advance nor can detailed routines for their accomplishment be specified. Long-term considerations are important, and a balance has to be struck between the accumulation of technological capital and the satisfaction of immediate commercial needs. The professional aspirations of a highly qualified staff must be satisfied while at the same time effective project management must be maintained. Dangers of remoteness and alienation from the rest of the organization were also noted.

Existing organizational structures were evaluated in relation to the criteria previously identified. The conclusions from this examination are summarized in Table 7.1.

None of these organizational structures satisfies all the criteria. New highly flexible organizational forms will be needed in the future, possibly combining several of the structures into a hybrid or multistructure pattern of organization.

In conclusion, it must be stressed that an organizational structure only provides a framework; it cannot guarantee the achievement of technological innovation.

> In the last analysis the quality of endeavour depends upon the quality of the people involved. No amount of organizational technique will make up for lack of integrity, intelligence, persistence, imagination, and the ability to help, enthuse, and understand one's fellows. Nevertheless, better organization should enable them to function more effectively.
>
> (F. Doyle, Former Research Director, The Beecham Group) [18]

We must not structure to the point where the organization encourages only

Table 7.1   *Characteristics of organizational structure*

	Degree of satisfaction of organizational criterion in the structure				
	Organiza-tion by discipline	*Project manage-ment*	*Product line or-ganization*	*Matrix organiza-tion*	*Venture manage-ment*
1. Development of technological capital	High	Medium	Low to medium	Medium	Low
2. Professional development of staff	High	Medium	Low to medium	Medium	Low
3. Managerial development of staff	Low	Medium	Medium	High	Very high
4. Achievement of short-term project goals	Low	Medium	Medium to high	Medium to high	Very high
5. Involvement of marketing, production and financial staff	Low	Low	Medium	Medium to high	High
6. Technology transfer	High	Medium	Low to medium	Medium	Low
7. Corporate identification	Low	Low	Medium	Medium	Medium to high

what it's already doing – but better. It must also encourage creation of new opportunities and there must be room for individuals who don't fit the structure.

(Sir Alastair Pilkington FRS) [19]

References

1. Burns, T. and Stalker, G. M. *Management Innovation*, Tavistock Publications, 1961.
2. Lawrence, P. R. and Lorsch, J. W. *Organization and Environment*, Division of Research, Graduate School of Business, Harvard University, 1967.
   —'Organizing for Product Innovation', *Harvard Business Review*, Nov./Dec. 1964.
   —*Developing Organizations – Diagnosis and Action*, Addison Wesley, 1969.
3. Lippitt, Gordon L. *Organization Renewal*, Appleton-Century-Crofts, 1969.
4. Barnes, Louis B. *Organizational Systems and Engineering Groups*.

Division of Research, Harvard Graduate School of Business Administration, 1960. Reprinted in: *Administering Research and Development*, Orth, Bailey, and Wolek (eds), Tavistock Publications, 1965.

5. Shepard, Herbert A. 'The Dual Hierarchy in Research', *Research Management*, Autumn 1958.

6. Gunz, H. P. 'Dual Ladders in Research: A Paradoxical Organisational Fix', *R & D Management*, **10**, No. 3, June 1980.

7. Baumgartel, Howard. 'Leadership Style as a Variable in Research Administration', *Administrative Science Quarterly*, Dec. 1957.

8. Ansoff, H. I., and Stewart, J. M. 'Strategies for a Technology-based Business', *Harvard Business Review*, Nov./Dec. 1967.

9. Twiss, B. C. and Goodridge, M. *Managing Technology for Competitive Advantage*, Pitman, 1989.

10. Mansfield, Edwin, *et al. Research and Innovation in the Modern Corporation*, Macmillan, 1972.

11. Burns, T. 'Research, Development and Production: Problems of conflict and cooperation', *IRE Transactions on Engineering Management*, Mar. 1961.

12. Goodridge, M. and Twiss, B.C. *Management Development and Technological Innovation in Japan*, Manpower Services Commission, Report No. MC28, 1986.

13. Howells, J. 'The Location and Organization of R & D', *Research Policy*, **19**, No. 2, Apr. 1990.

14. Belbin, R. M. *Management Teams: Why They Succeed or Fail*, Heinemann, 1986.

15. Barczak, G. and Wileman, D. 'Leadership Difference in New Product Development', *Journal of Product Innovation Management*, **6** No. 4, Dec. 1989.

16. Larson, E. W. and Gobeli, D. H. 'Organizing for Product Development Programmes', *Journal of Product Innovation Management*, **5**, No. 3, Sept. 1988.

17. Katz, R. and Allen, T. J. 'Investigating the "Not Invented Here" Syndrome', *R & D Management*, **12**, No. 1, Jan. 1982.

18. Doyle, F. P. *A Fresh Look at the Management of Large Scale Creative Research*, Address to the R & D Society, Jan. 1971.

19. Pilkington, Sir A. 'Science and Technology', *Proceedings of the Royal Society of Arts*, **CXXIV**, No. 5241, Aug. 1976.

Additional references

Alsop, K. 'Managing Scientists and Technologists', in *Technology and Management*, Wild, R. (ed.), Cassell, 1990.

Goold, M. C. and Campbell, A. *Strategies and Styles*, Blackwell, 1987.

Handy, C. *Understanding Organisations*, Penguin, 1986.

Hanan, M. 'Corporate Growth Through Venture Management', *Harvard Business Review*, Jan./Feb. 1969.

Houghton, J. R. 'The Role of Technology in Restructuring a Company', *Research Management*, **XXVI**, No. 6, Nov.–Dec. 1983.

Kingdon, D. R. *Matrix Organisation: Managing Information Technologies*, Tavistock Publications, 1973.

Knight, K. (ed.) *Matrix Management: A Cross-functional Approach to Organisation*, Gower Press, 1977.

Peters, T. J. *Thriving on Chaos*, Macmillan, 1988.

Roberts, E. B. 'New Ventures for Corporate Growth', *Harvard Business Review*, July 1980.

Roberts, E. B. and Fusfield, A. R. 'Staffing the Innovative Technology-Based Organization', *Sloan Management Review*, Spring 1981. Reprinted in *Generating Technological Innovation*, Oxford University Press, 1987.

Roth, L. M. 'A Critical Examination of the Dual Ladder Approach to Career Development', *Columbia Journal of World Business*, 1982. Reprinted in *Readings in the Management of Innovation*, 2nd edn. Tushman, M. L. and Moore, W. L. (eds), Ballinger Publishing, 1988.

8

Technology forecasting
for decision-making

Thus in policy research we are not only concerned with anticipating future events and attempting to make the desirable ones more likely and the undesirable less likely. We are also trying to put policymakers in a position to deal with whatever future actually arises, to be able to alleviate the bad and exploit the good.

Herman Kahn

The need to forecast

All we know for certain about the future is that it will be different from the present. Thus the products, organizations, skills and attitudes which today serve a business well may have little relevance to the conditions of tomorrow. If a business is to survive it must change. And the changes must be timely and appropriate to meet the needs of the future. The difficulty arises in prognosticating what these needs will be. Nevertheless, an attempt must be made to do so and, however imperfect forecasts may be, managers cannot afford to take decisions affecting the future of their organization without examining any clues they can find to the best of their ability.

Forecasts are important inputs to the process of strategy formulation and planning. They have been used to gain a better understanding of the threats and opportunities likely to be faced by established products and markets and, consequently, of the nature and magnitude of the changes needed. Thus, anticipation enables the business to be steered into the future in a purposeful fashion, in contrast to belated reaction to critical events. Nowadays, technical lead times are often so long that a market can be lost before a proper response is made. While this approach may be satisfactory conceptually it is of little value unless one can make sufficiently accurate forecasts to aid the practical manager in his decision-making. Numerous techniques for technological forecasting have been developed with the object of enabling the manager to obtain the maximum use from the information available to him.

Since technology is responsible for many of the most important changes in our society, forecasting future advances in technology and their impact can be as vital for top management in the formulation of corporate strategy as it is for the technologist reviewing his R & D programme. For technological changes may sometimes result in a major redefinition of an industry or market. We have seen for example, that the producer of metal cans may find his major threats coming not from direct competitors in can manufacture, but from other packaging technologies such as glass, plastics or paper, or from different forms of preservation such as freeze drying. However, the area of consideration might need to be widened even further to take in factors such as: the size of the future market for convenience goods upon which the relative economic merits of alternative packaging technologies may depend; the availability and future costs of packaging materials; customer tastes in convenience products; health and hygiene standards; and a multitude of similar items. Similarly, the manufacturer of lawn mowers may eventually suffer more dangerous competition from the development of grass retardants by the chemical industry. But in order to assess this threat it is necessary, not only to forecast the technological possibility of a satisfactory product being developed; but also to predict whether it would be socially acceptable to the market.

It follows that the field of investigation has to be very extensive if some important indication is not to be missed and it must include a wide range of social, economic, environmental and political as well as technological factors. While sometimes identification of the threat or opportunity may be sufficient, more frequently the real implication can only be properly evaluated by a detailed study. When it is realized that each factor is associated with a high degree of uncertainty, the task of the forecaster might seem wellnigh impossible. Yet, however daunting the difficulties may appear, the manager cannot avoid making decisions which will be proved good or bad by future events. He must take a view of the future. If forecasting techniques can enable the manager to obtain a more accurate picture of the future and in consequence, improve his decision-making, the effort devoted to it is justified. This can be the only real justification for forecasting.

In any consideration of forecasting, it is important not to lose sight of the high degree of uncertainty in the outcome. The future will never be predictable. Forecasting can assist in the formation of managerial judgment, but will never replace it. Many of the critics of forecasting condemn it for its failures, of which there are abundant examples. Perhaps the forecasters themselves are largely to blame for encouraging others to place more reliance on their forecasts than they merit. We must learn to accept that many forecasts will remain poor, but appreciate that if conducted conscientiously, they will lead to fewer errors and the avoidance of some of the most costly mistakes.

With this perspective in mind, it can be seen that there are limitations which must set some bounds to the resources devoted to forecasting. For like any other managerial activity, the investment of time and resources can only be justified in terms of the benefit expected to derive from it. This criterion must be used to prevent excessive expenditure on forecasts which do not aid decision-making, however interesting they may be in themselves. As Drucker [1] comments:

Decisions exist only in the present. The question that faces the long range planner is not what we should do tomorrow, it is: What do we have to do today to be ready for an uncertain tomorrow? The question is not what will happen in the future. It is what futuristics do we have to factor into our present thinking and doing; what time spans do we have to consider, and how do we converge them to a simultaneous decision in the present?

In some industries today's decisions cover a long time-scale. This is true of fuel, power generation, and communications where the effects of what is decided now will still be felt several decades hence. In the case of a power station, for example in the choice of the reactor system for a nuclear plant, the design and construction stages may take five to six years to be followed by an operating life of twenty years or more. Although the importance of the more distant dates loses significance because of financial discounting it may still be necessary to make forecasts for a period of twenty or more years.

By contrast, the planning horizons for most companies are much shorter — of the order of five to ten years. It is still important to forecast, for many significant changes can occur in a decade. Thus the only meaningful determinant of the time period for which forecasts are necessary would appear to be the planning horizon of the company. This is a function of the rate at which the company's activities can be made to respond to changes rather than the rate at which the environment itself is changing.

Summarizing, we can see that forecasting can assist business decision-making in the following ways:

1. Wide ranging surveillance of the total environment to identify developments, both within and outside the business's normal sphere of activity, which could influence the industry's future and, in particular, the company's own products and markets.
2. Estimating the timescale for important events in relation to the company's decision-making and planning horizons. This gives an indication of the urgency for action.
3. The provision of more refined information following a detailed forecast in cases where an initial analysis finds evidence of the possibility of a major threat or opportunity in the near future, but where this evidence is insufficient to justify action,
   or
   continued monitoring of trends which, while not expected to lead to the necessity for immediate action, are, nevertheless, likely to become important at some time in the future and must consequently be kept under review.
4. Major reorientation of company policy to avoid situations which appear to pose a threat or to seek new opportunities by:
   (a) Redefinition of the industry or the company's business objectives in the light of new technological competition.

(b) Modification of the corporate strategy.

(c) Modification of the R & D strategy.

5. Improving operational decision-making, particularly in relation to:

(a) The R & D portfolio.

(b) R & D project selection.

(c) Resource allocation between technologies.

(d) Investment in plant and equipment, including laboratory equipment.

(e) Recruitment policy.

Level of investment in technology forecasting

There is a common assumption that technology forecasting (TF) is for the larger companies, since it is only they who can afford to undertake the exhaustive process of data collection and analysis required by the more sophisticated techniques. Technological forecasters often point to the dangers of forecasts based on inadequate analysis and argue that useful results can only be expected when information is meticulously gathered and carefully analysed.

TF then, appears to be an expensive activity employing a multidisciplinary team in detailed analysis. Without doubt, effort on this scale is normally essential if the most accurate forecasts are to be prepared. But this is a counsel of perfection. Small companies equally face an uncertain future and need to build some form of forecasting into their decision-making even if it is based only on the judgment of the chief executive. If his judgment can be assisted by simple forecasting techniques, then they should be used, even though it is realized that more accurate forecasts might have resulted from the investment of greater effort in sophisticated techniques. The question should not be whether or not to forecast, but:

1. To what extent is it necessary to forecast?

2. Which are the most appropriate techniques to use, given the limitation of what can be afforded?

When attention is focused on need rather that the ability to pay, it can be seen that the importance of forecasting is much more a matter of the business environment than the size of the company. The large firm in a mature industry is unlikely to be overtaken suddenly by a technological development which will have a catastrophic effect. For such a company TF can be confined to monitoring the environment to give early warning of the occasional advance; only then does it need more detailed forecasts. It was not, for example, St Gobain's ignorance of the float glass process which led to its late investment in the new technology, but its lack of adequate response to the threat. TF might have helped the company's understanding but its effect upon the decision-making process is likely to have been negligible.

By contrast, one can see how TF could have been applied both by Rolls-Royce and its competitors when the Trent engine was being evaluated. Aero-engine technology was advancing and forecasts showed a very close correlation between an engine's performance and the date of its introduction. The rate of technological change in an industry should, therefore, be a much better criterion than size of company as a determinant of the need for TF, although as we have seen earlier, the speed at which a company can react to change is also important.

The obscurity of the precise nature of a threat also has an important bearing on the degree of detail required in a TF. Sometimes the threat is obvious, as in the case of the manufacturer of lead additives for petrol – the only area of doubt is in the timescale of anti-pollution legislation. On the other hand, the maker of lead acid batteries has a much more complex problem to resolve. First, he has to consider whether the conventional reciprocating petrol engine will be totally banned, or whether the petroleum or motor industry will find an acceptable solution to exhaust pollution. If banned, it might be replaced by the Wankel engine, gas turbine, steam or electric propulsion. Even if his analysis shows a high probability of electric traction being widely adopted, he will be aware that major motor manufacturers are engaged in research into new forms of electric power storage. It is, therefore, not immediately obvious whether he is facing a threat or an opportunity. In this situation, detailed technology forecasting could be of invaluable assistance.

An offensive R & D strategy exposes a company to much greater risks than a defensive strategy, particularly in relation to the state of the technological art and timing. Greater steps into the unknown are involved. In this situation, opportunities are being sought, rather than the avoidance of threats. Thus a company adopting an offensive strategy should devote correspondingly more resources to TF.

In summary, we can conclude that:

1. All companies should undertake some form of technology forecasting.
2. The amount of effort devoted to TF should take into account:
   (a) The rate of change in the environment.
   (b) The planning horizon determined by the technological and marketing lead times for new products or processes.
   (c) The complexity of the underlying problems.
   (d) The R & D strategy.
   (e) The size of the company only in so far as the availability of resources limits the choice of techniques which can be afforded.

What drives technological progress?

Technological progress does not occur naturally; it is made to happen. If the forces driving it can be identified and quantified it should, therefore, be possible to forecast the rate at which they will produce future advances. The direct cause is the work of individual technologists or teams which would not happen in the absence of financial support. Thus one would expect to find a relationship between the level of funding and technological progress (Twiss [2], Foster [3]).

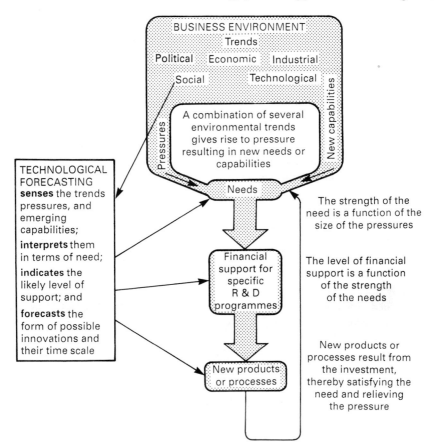

8.1 Technological forecasting and the innovation chain

The funding of technology occurs because the investor considers that the benefit to be derived from the enhanced performance is greater than the cost of achieving it. This benefit is gained in the market for a commercial development or from the broader needs of society for a government-funded project. Thus it can be seen that technology is driven by market needs or a perception of what these needs may be in the future. This leads to the important conclusion that any forecasting activity must be based upon an understanding of these market forces. These, in turn, depend upon a complex interaction of a host of economic, social, political and environmental factors as well as the potential of technology to satisfy them. Figure 8.1 illustrates these causal relationships which, in theory, could form the basis for technology forecasting. Examination of the figure suggests two ways in which this might be done:

1. *The development of a mathematical model.* This would be analogous with an econometric model which identifies all the significant factors, how they interrelate and quantifies these relationships. A computer program could then

be written which enables forecasts for all the factors to be combined to produce the forecast for the technology being investigated.

2. *Examination of the relationship between investment and progress.* The cumulative past investment in a technology and the progress it produced could be plotted on a graph showing the trend which could then be extrapolated into the future to provide a forecast.

Attractive as these approaches may appear in theory, there are a number of reasons why they cannot be used in practice. Thus when one attempts to construct a practical model it is found to be impossible because of the extreme complexity of the large number of factors to be incorporated, the inability to quantify many of the influences such as social and political factors, and the absence of data to enable the linkages to be quantified. Nevertheless, study of the model without attempting a quantification is a useful aid to the forecaster in enabling him to understand the underlying influences shaping his technology.

Difficulties also arise in establishing the relationship between the cumulative investment and technological progress. In most situations this information is normally not available, although it should be used in those cases where it can be obtained. Even so, it does not lead directly to what the forecaster usually requires in order to make his decisions — the relationship between progress and time.

If these rigorous causal methods cannot be applied other approaches must be sought. These are based upon:

3. *Examination of the relationship between technological progress and time.* Because progress is not random, patterns can be established which relate past progress with time. These can then be extrapolated into the future even when the processes which have produced these advances are imperfectly understood. Indeed examination of these trend patterns may suggest that future progress is inevitable. This is not so. If the factors creating the market need change, so will the rate of financial support. If for any reason this support is withdrawn, progress will cease; this may be because of the emergence of a competitive technology or because the current level of performance fulfils the needs of the market.

Thus the practical techniques of forecasting are not based upon the explicit use of causal models. Nevertheless, the forecaster must not lose sight of the driving forces behind the technology. He must understand the relationships, albeit at a judgmental level, and assess whether any changes might invalidate or modify the forecasts.

Inputs to the forecasting system

Like any other procedure for systematic analysis, forecasting can only be as accurate as the information fed into it. The inadequacy of the input data cannot be compensated, though it might easily be hidden, by highly sophisticated quantitative analysis. And what we have at our disposal is limited to:

1. Information from the past.
2. Knowledge of the present.
3. The ability of the human intellect – logical thought process, insight and judgment.

The collection of data is often the most difficult task at the outset of a forecasting activity. In many technologies there has been no systematic record of what levels of technological performance have been achieved at dates in the past. Considerable research may be required to uncover this information. Even so, it may not be very precise.

The relatively poor quality of this data obviously affects the accuracy of the forecast based upon it. This is a real difficulty and may appear an inherent weakness of TF. However, the purpose of TF must not be forgotten. It is an aid to decision-making. For the most part long-term decisions are seeking answers to *big* questions such as: 'Do we or do we not invest in this new technology?' The rationale for TF is that the systematic analysis of the best available data using the best minds will lead to a *better* decision than one based solely on judgment or intuition. Furthermore, as experience is gained in TF the quality of the data will improve due to the recording of contemporary and more accurate data.

Assumptions are in effect high probability forecasts which can be taken as givens in the subsequent analysis. However, it is important to record the assumptions and review them periodically to ensure that they remain valid.

These are the only resources to be marshalled and interpreted within a forecasting system. Discussion of TF inevitably centres on the detail of the techniques themselves, but this must not be allowed to distract attention from the two factors that determine the usefulness of the results – the quality of the input data, and the calibre of the minds applied to the task.

Outputs from the forecasting system

There are four essential elements of a comprehensive forecast – qualitative, quantitative, time and probability (Fig. 8.2).

The qualitative element relates to the phenomenon or event to be forecast (*What should I forecast?*). This is the area where insights and the general technological awareness of the forecaster play a critical part. It is of little avail to conduct detailed studies of a mature technology such as precision engineering when the future lies with electronics. In this case the precision engineer must not only be aware of developments in electronics but also be able to assess its impact upon his company. This will provide the basis for formulating a proactive R & D policy.

Once the phenomenon to be studied has been identified it is essential to

INPUTS   FORECASTING METHODOLOGY   OUTPUTS

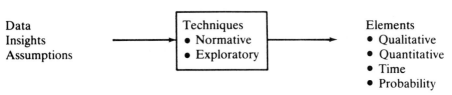

Data   Elements
Insights   • Qualitative
Assumptions   • Quantitative
                                                              • Time
                                                              • Probability

8.2 Inputs and outputs from forecasting

determine a measure for it (*What data do I need?*). As we shall see later the choice of the appropriate attributes and parameters is critical. The parameter allows the plotting of data from the past and its extrapolation into the future.

Forecasts without quantification and timescales are largely useless (*What rate of progress can I expect?*). It is often relatively easy to forecast the eventual development of a technology to produce an end state scenario, but it is much more difficult to assess the path by which this end state will be achieved. Many detailed techno-economic analyses often lack this critical element. Without a timescale R & D planning is of little value.

Since forecasts are concerned with the future they are uncertain; they need to be associated with a probability assessment (*What confidence do I have in my forecast?*). It should be noted, however, that there are some events likely to have a profound impact on a technology which, although foreseeable, cannot be related to a probabilistic estimate of a timescale. For example, the possibility of an accident in a nuclear power station could be foreseen but no realistic forecast could be made of when the incident might occur (e.g. Chernobyl). Nevertheless, an assessment of the likely consequences could be made in advance and should form part of the forecasting process.

Any forecast which does not contain all four elements cannot provide an adequate basis for decision-making. Thus a statement: 'By the year 2000 solar heating could be widely used in the UK' is of little value whereas 'Solar heating (qualitative) will provide 20% of UK domestic energy (quantitative) by the year 2000 (time) with an 80% probability (probability)' satisfies all the criteria.

The use of the words *could* and *will* in these two statements is worthy of consideration. The application of the TF techniques is likely to lead first to a forecast of what *could* happen whereas the manager must base his decision on an assessment of what *will* happen. Obviously the event will not occur unless it can happen but the fact that it could occur does not necessarily mean that it will − the '*wills*' are a subset of the '*coulds*'. Having established a feasibility it is then essential to consider the processes whereby the potential might be translated into an actuality. This will usually involve consideration of a variety of economic, political and social factors which are often difficult to quantify. Failure to appreciate this important distinction and to build all these factors into a forecasting output is a common cause of over-optimistic forecasts leading to poor managerial decisions. The reader might like to consider the solar heating example from which it might be concluded that the actual level is likely to be significantly lower than it could be.

Classification of forecasting techniques

The techniques of TF are commonly classified under the headings of 'exploratory' or 'normative'. The term exploratory covers those techniques based upon an extension of the past through the present and into the future. They look forward from today taking into account the dynamic progression which brought us to today's position.

A normative approach starts from the future. The mind is projected forward by postulating a desired or possible state of events represented by the accomplishment of a particular mission, the satisfaction of a need, or state of technological development. It is then necessary to trace backwards to determine the steps necessary to reach the end-point and assess the probability of their achievement. In the elaborate scenarios developed by the research of 'think tanks' alternative futures for the whole of mankind are developed. These workers claim that the scenarios present a choice from which mankind can select the most desirable of the alternative futures enabling the formulation of policies and the allocation of resources to travel the path to reach the desired end. However, these prognostications are beyond the concern of the industrial manager whose objectives are specific and limited. Nevertheless, they do underline that there is always a choice to be made.

In a practical forecasting exercise, it is usually necessary to employ a combination of techniques, some of which may be exploratory and some normative. For in dealing with great uncertainty, it is essential to examine every clue from as many angles as possible.

Too much stress must not be placed on 'making the future happen'. It is rare that one company or series of decisions has a profound influence on the future of technology, although the consequences of a decision may be of great significance for the company concerned. If one firm does not proceed with a potential innovation there is a high probability that another will do so within a short space of time. The history of technology contains many examples where similar innovations occurred almost simultaneously in several places. This is not chance, but the result of a combination of the advances in several technologies necessary for the achievement of an innovation. It was not, for example, the discovery of the principle of the gas turbine, which had been known for many years, that made possible its practical realization in the late 1930s, but the availability of high temperature materials. Project Hindsight provides further evidence and concludes, 'Engineering design of improved military weapon systems consists primarily of skilfully selecting and integrating a large number of innovations from diverse technological areas so as to produce systematically, the high performance achieved.' This statement is supported by evidence from several examples investigated during the research programme:

As an example, transistor technology is credited with being responsible for size reduction in electronic equipment, however, without the development of such ancillary technologies as tantalytic capacitors, high core permeability chokes and transformers, printed circuits, dip soldering, nickel cadmium batteries and silicon cell power supplies, the electronic equipment chassis would be only marginally (perhaps 10%) smaller than a vacuum tube version.[4]

Thus, if a forecasting exercise is to give useful guidance to decision-makers regarding a specific innovation, it must take into account not only a wide range of technological advances but also their mutual interactions. Furthermore, timing is of the essence. Before a certain date the convergence of technological capabilities is insufficient to support the innovation. Once that stage had been reached, competitive forces are likely to ensure a limited time advantage to the company that seizes the initiative. TF can assist in deciding when to start. But the competitive edge so gained will be quickly eroded by an R & D programme poorly conceived or conducted without vigour.

Most businesses are living in the type of environment we have just examined, where opportunities will emerge from marginal improvements in a number of advancing and converging technologies. In this situation exploratory techniques of TF are likely to be the most helpful.

Mission-orientated programmes are rarer. Frequently they result from socially or politically motivated needs, and are likely to involve major investments. The end-point of the project will be postulated − landing a man on the moon, low toxicity motor car exhausts, a cancer cure, or an antiballistic missile system. Normative techniques, usually in conjunction with exploratory methods, can be of assistance here in forecasting the likely time of achieving the desired outcome with different levels of support, evaluating the alternative paths and selecting the best, and estimating the probabilities of each component of the overall programme being satisfied within a given timescale.

The paths of technological progress

If technological progress consisted of a succession of random events any attempt to forecast would be impossible. Fortunately, however, analysis of historical data from a considerable number of phenomena shows that progress is not random and discontinuous, but follows a regular pattern when a selected attribute, such as functional performance (e.g. aircraft speed), a technical parameter (e.g. tensile strength to density ratio for a material), or economic performance (e.g. cost per kWh for electrical generation) is plotted against time. Characteristically one finds an S-curve pattern (Fig. 8.3).

The growth of a total market exhibits the same S-shaped characteristic although the limiting level in this case is set by market saturation which must itself be forecast (Fig. 8.4). These curves which are specific to one technology or market, are measurable and can be plotted on a graph. They should not be confused with the industry life-cycle which is more conceptual and provides

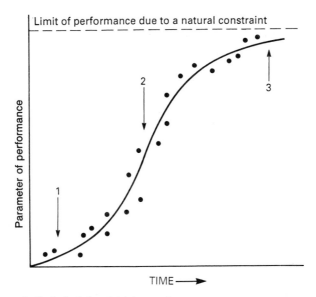

1. Period of slow initial growth.
2. Rapid exponential growth.
3. Growth slows as performance approaches a
   natural physical limit asymptotically.

8.3 The S-curve of technological growth

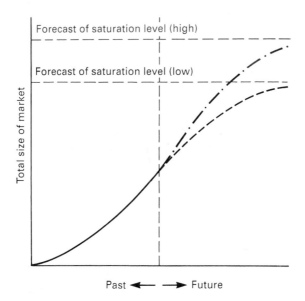

8.4 The S-curve of market growth

an illustration of its evolution rather than an accurate representation. Nor is it the same as a product life-cycle which relates to one clearly defined model. These differences can be seen in relation to the motor car:

1. *Technological S-curves* – a number of technological parameters of which specific fuel consumption would be one example.
2. *Market S-curve* – the growth of the market represented by the parameter cars per head of population.
3. *Industry life-cycle* – the growth of the motor industry which is more difficult to determine since it is dependent upon many factors such as car ownership levels, the value added per car, average annual usage and the car life expectancy.
4. *Product life-cycle* – this relates to a specific model which has a limited life before replacement by another model which may embody little new technology.

In this chapter we are concerned with only 1 and 2.

Two important guidelines for R & D management can be inferred from observation of the shape of the S-curve.

1. The human intellect, with its linear thought patterns, may seriously underestimate the rapidity of the potential progress when establishing design specifications during the exponential mid-phase.
2. The decreasing managerial returns which may be expected from investment in a technology as its physical limit is approached.

It has been seen that there is no direct causal relationship between progress and time. The justification for drawing an S-curve and using it for forecasting is based upon historical evidence covering a large number of technological parameters which shows that they have almost invariably followed this shaped path. There is thus a high level of confidence that a new technology for which only part of the curve is known will progress towards a natural limit on a S (logistic) curve. But it must be recognized that this is a consequence of human behaviour which, although the evidence is circumstantial, leads to a regular rate of evolution. This permits an extrapolation unless some factor which will modify that behaviour can be identified. How such changes might affect the curve are shown in Fig. 8.5. If the stimulus to invest in reaching a higher performance is low, the curve described is likely to be of the form OB. By contrast, OC_2 exhibits the much greater rate of progress to be found when market needs have led to a high expenditure in R & D. It is also possible to modify the shape of the curve; for example, a technology having reached the point A_1 on curve OA, at time t_1 may receive a stimulus from a needs-oriented mission which causes it to follow the path $A_1C_1C_2$; there is, of course, a limit to the maximum slope of the curve set by the size of the resources available or the rate at which they can be usefully employed. From the viewpoint of a forecaster at time t_1

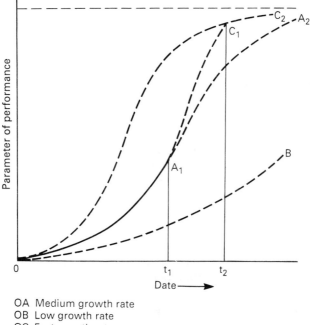

OA Medium growth rate
OB Low growth rate
OC Fast growth rate

The rate of growth is largely determined by the technological effort devoted to it. A technology which has advanced along OA_1 to time t_1 may be expected to progress along $A_1 A$ *unless* an identifiable outside factor causes an acceleration along $A_1 C_1$

8.5 Possible paths of progress

attempting to predict the future with the knowledge that progress to date has followed the path OA_2 the indications are that the curve A_1A_2 has a high probability of describing the immediate future, unless he can positively identify some factor likely to cause a discontinuity producing a departure along a different path. In practice, such occurrences are infrequent.

When considering a specific technological development the forecaster will not have available to him the eventual shape of the S-curve. Much of it will lie in the future and he will have to take a view of the following.

The period before growth becomes rapid

The initial stage of low growth can often extend for many years. Over-optimism frequently leads to over-investment even where the technology eventually reaches the stage where it diffuses rapidly. There are many such cases – the videophone, an early example of which AT&T exhibited at the 1964 World Fair; carbon fibres, where the promise of the 1960s is beginning to be fulfilled only in the 1990s as the material gains acceptance in aircraft construction; and the automatic factory widely discussed in the 1950s which is only now emerging with the growth of robotics. All these misjudgments resulted from an incorrect assessment of the timing rather than of the ultimate potential of the technology.

The slope of the curve during rapid growth

This slope is influenced by several factors, the most important of which are:

1. The advantage in price, reliability or performance offered by the new technology which determines its appeal to potential customers. Thus the electronic calculator gained market rapidly.
2. The level of investment required, which can discourage the scrapping of old plant in a mature industry, for example float glass or basic oxygen steel.
3. The infrastructure investment necessary to exploit the new technology fully, a limiting factor in many electronic developments such as cable and satellite TV or facsimile transmission.
4. The unexploited potential of the existing technology which may respond to the new competition, as seen in the development of thin glass to counter the threat from the plastic PET bottle.

When growth will begin to slow appreciably

As an industry or technology approaches its upper limit there are psychological barriers to accepting that the period of exponential growth is coming to an end. At the industrial level this can lead to over-capacity as we have seen in recent years in ethylene plants, oil refining and motor manufacture. In R & D the response is often to invest heavily in projects based upon the existing technology where only incremental improvements can be expected, rather than to seek the opportunities offered by other, often new, technologies which could provide better commercial prospects.

An approaching physical limit does not remove the need to forecast. For it is at such a time that a new technology may emerge. This new technology will have a different physical limit and a potential for further progress in performance. Eventually it is likely to supplant the existing technology in a wide range of applications, particularly when they call for high performance.

We can see that study of the past progress of technology confirms the existence of regular patterns, and provides a framework within which forecasting can be undertaken. It also indicates the types of information that the R & D manager would like to extract from forecasts as:

1. The performance levels likely to be reached within his decision-making horizons in the absence of any new stimuli.
2. The identification of stimuli which may change the level of technological support and the effect this would have on the rate of progress.
3. The approach of a natural or physical limit for a technological parameter and its impact on the rate of progress.

4. The emergence of new technologies and the identification of cross-over points:
   (a) without additional investment in the established technology;
   (b) with changed levels of investment.
5. The determination of performance milestones in several technologies the achievement of which would together make feasible a specific innovation.

The techniques of technology forecasting

It is beyond the scope of this book to explain in detail, the many techniques for technology forecasting now available and fully described in the literature [5, 6, 7]. Table 8.1 lists a number of the techniques of technology forecasting. The following brief descriptions of a few of the most widely used techniques will not attempt to provide the reader with a working basis for their practical application; they are intended to show how the principles described previously can be applied and some of the major problems likely to be encountered. But it is worth reiterating that the techniques are not an end in themselves, and their successful application must rest heavily upon the technological experience and insights of the managers using them.

Table 8.1 *The techniques of technology forecasting*

Qualitative methods
Creativity spurring techniques – brainstorming, synectics, lateral thinking
Time independent contextual mapping
Analogies
Morphological analysis
Gap analysis
Environmental surveillance and monitoring
Scenarios

Quantitative and time forecasting methods
Attribute and parameter identification
Time series – growth and logistic curves
Envelope curves
Precursor events and curves
Substitution curves
Relevance trees
Quantified analogies
Trend impact analysis
Dynamic modelling

Probability forecasting methods
Delphi
Cross-impact analysis
Gaming methods

The use of the S-curve

Simple trend extrapolation
Until the S-curve has begun to manifest itself there is no way of obtaining the

shape of the progress of a new technology based upon data. Some form of judgmental forecasting such as Delphi (see p. 287) must then be used. This is not normally, however, a major problem since most TF activity relates to a technology which has already established a presence on the S-curve.

The first decision facing the forecaster is the selection of the appropriate parameter to plot. This is not so obvious a choice as may at first appear. It is necessary to go back to the market and determine what attribute the purchaser of a product desires. In some products there may be one dominant attribute but in others there may be a number, each of different importance for the various market segments. For example, in the motor car these might include speed, economy, road holding or comfort. It is then necessary to identify the technical parameter relevant to the appropriate attribute for the market segment the product is designed to satisfy. This parameter may not necessarily be the one on which the industry's technologists naturally focus.

A classical example of this occurred when jet engines were first considered for commercial aviation. Aero-engine designers had traditionally regarded specific fuel consumption (sfc) as the dominant design criterion. This remained appropriate so long as there were no major changes to the aircraft characteristics. Some, therefore, initially rejected the jet engine because of its high sfc. However, consideration of the market showed that the attribute of importance to the airline operator was economy of operation represented as cost per passenger (or tonne) kilometre. The increased productivity of the jet aircraft due to its higher speed more than outweighed the increased fuel consumption and made it a viable airliner. A more appropriate measure for the engine manufacturer would be the fuel consumption per passenger (or tonne) mile. This shows that the attribute required of the total system must be considered and not just one component of it. In the jet airliner case this remained unchanged but the implications for the contribution of the engine to this was overlooked. Technologists obsessed by their parochial concerns can lose sight of this and be led into faulty decisions.

Forecasting the wrong parameter can lead to bad technical decisions. This can only be avoided if the needs of the user of the total system and the attributes he requires are carefully considered before embarking on detailed forecasting.

Once the parameter has been selected it is necessary to obtain data for its past progress over time. This can then be plotted to form part of an S-curve. It then becomes necessary to extrapolate this in order to provide the forecast for future performance levels. One way of doing this is to apply one of the standard curve fitting techniques which when extrapolated would become asymptotic at the top of the curve. This is often done but is likely to lead to a poor forecast for two reasons. First, the inaccuracies inevitable in the data are likely to yield a curve which does not become asymptotic at the correct level. Secondly, it does not make use of the knowledge of the natural physical level which can be determined and should be used. Thus the recommended procedure is to fit a logistic curve to the data which is also asymptotic to that level. Once drawn the implications for technical planning can be evaluated.

A systematic approach to the preparation and use of the forecast is as follows:

1. What is/are the key attribute/s the market requires from the system in which the product is embodied?
2. What technical parameter is relevant to the satisfaction of the chosen attribute?
3. How can this parameter be measured?
4. Where can this data be obtained?
5. What is the theoretical limiting level of this parameter using current technology?
6. What factors may cause future progress to depart from the S-curve obtained by plotting this data?
7. How might these factors modify the shape of the curve?

At this stage the implications for technical planning can be addressed as follows:

8. What is the gap between the current level of achievement and the theoretical limit?
9. What is the significance of this gap for product/technology policy?
10. Can an existing or potential technology with a higher limiting level be identified? If so, what are its implications for the R & D programme?

The S-curve as described is of particular value when the company's product has been dependent upon one technology in the past and is likely to remain so for the future. This is a pattern of steady evolution and gives a strong indication of the level of performance which should be incorporated in the specification for a new product.

Multiple parameters
Many products have a number of attributes all of which have to be satisfied to some extent. The relative importance attached to them varies with the market segment. However, the interrelationship between technologies limits the extent to which they can all be satisfied. There must be tradeoffs, for example between speed and economy for the motor car. At any time it is possible to postulate a 'state of technology' combining the contributions of several technologies where these tradeoffs exist. Thus it is possible to represent the state of technology (T) as:

$$T = f(\text{Parameter A})^x \times (\text{Parameter B})^y \times (\text{Parameter C})^z$$

where x, y and z are indices which can be derived from the data. When T is plotted it is found that this also follows an S-curve. For example, it has been shown that armoured fighting vehicles, comprising tanks, armoured cars and self-propelled guns, fall on the same curve when the attributes of speed, defensiveness and offensiveness are combined using the parameters of kilometres per hour, armour thickness and gun calibre respectively. This proved adequate from 1916 until the late 1970s when the emergence of radically new technologies such as Chobham and explosive armour invalidated the use of armour thickness as a measure for defensiveness.

S-curves and technical change

One of the most valuable contributions of TF is the identification of those times when a major change in a company's technological emphasis may be required. This will now be examined in relation to:

1. Technology substitution.
2. Attribute substitution.
3. Product substitution.

Technology substitution

A clear distinction must be made between a technological parameter and the technology by which it is achieved. A new technology, if it is to be successful, will normally have a higher natural limit than the old when measured by the same parameter, thereby permitting further progress to be made. In some cases, however, the new technology may enable a natural limit to be approached more quickly or with a lower development cost or at a lower product price. The role of TF is to identify these technologies and assess their potential to aid in deciding when to withdraw support from the old in favour of the new.

This can be illustrated by considering the internal combustion engine. Thermodynamic efficiency, an important parameter in engine design, is dependent upon combustion temperature which is constrained by the physical properties of the material from which it is manufactured. This is currently approaching a limit with economically obtainable metals. However, ceramics are able to sustain higher working temperatures and a successful ceramic engine would achieve a thermodynamic efficiency impossible with metal. This is the rationale for the current research into the ceramic engine, particularly in Japan. There are, of course, many practical problems to be overcome but the technological logic is clear; once these problems are solved economically the metal engine will be unable to compete (Fig. 8.6(a)). Similarly, in many applications speed of operation is important. Thus mechanically operated systems, limited by inertial effects, have been replaced by electronics.

The reasoning is different in the case of time-keeping where the accuracy of mechanical devices had been improving along an S-curve for many centuries. Further development could have been expected to have moved it closer to the natural limit of zero error. The electronic watch, with the same limiting level, was able to approach it more quickly with the added benefit of lower product cost (Fig. 8.6(b)).

History shows that when a higher performance level is achievable with a new technology the practical problems are almost invariably overcome. There is thus a high probability that it will replace the existing technology at some time in

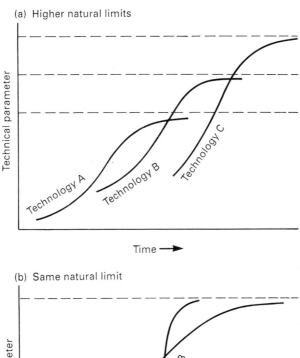

(a) Higher natural limits

Technical parameter

Technology A

Technology B

Technology C

Time →

(b) Same natural limit

Technical parameter

Technology A

Technology B

Time →

8.6 Technology substitution

the future. Comparison of the two S-curves gives a strong indication of when this cross-over will occur and assists the technologist in deciding when development effort should be transferred from the old to the new.

As would be expected from the S-curve, an emergent new technology at first grows slowly. This is because its early performance is likely to be inferior to the highly developed existing technology. But when the performance of the two technologies approaches the same level, the greater potential of the newcomer attracts increasing investment, particularly once it has taken the lead when it begins to grow exponentially.

No natural law governs the emergence of a new technology. There is a physical limit even to the discovery of new technologies as evidenced by the flattening of the 'envelope' curve. This phenomenon does, however, present the R & D manager with some particularly tricky judgments. The experience of history suggests that a slowing down or the approach of a natural limit should alert him to the possible appearance of something new. But his correct reaction to the identification of a substitute technology is not straightforward.

The initial expectation is that R & D effort will be transferred and that attempts to approach closer to the upper performance limit of the established technology will be reduced. However, in practice, the reaction may be the reverse. A great deal of capital is invested in the existing system. The threat may call for a defensive response and an increased rather than reduced investment when some development potential remains. This can be seen in Fig. 8.7 where an established technology (1) is threatened by progress in technology (2). Extrapolating the past trends suggests a cross-over point at time t_2 where T represents the present. But in spite of the approaching natural limit for (1) it is still possible to invest more heavily in it to follow the path TX rather than TY thereby delaying the cross-over point to time t_3. In the perspective of history the delay represented by $t_2 - t_3$ is insignificant, but it may be of great importance to the

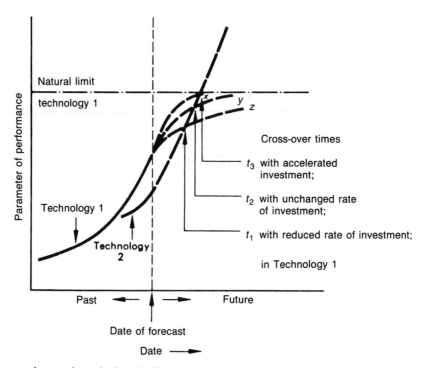

An emerging technology 2 is likely to replace the existing technology 1
The question is, when? What effect will the threat from 2 have on investment in 1?
And how will this influence the date of substitution of 2 for 1?

8.7 Effect of an emergent new technology on an established technology

short-term strategy of a business heavily invested in (1). This effect is often observed in practice.

This is often referred to as the 'sailing ship syndrome', reflecting the improvements in sailing ship performance in the 19th century when steam first began to be used for marine propulsion. There are several reasons for this:

1. The initial performance of products incorporating the new technology may be inferior to that obtainable with the existing technology; it is still in its early part of the S-curve.
2. The reliability of products using the new technology can be expected to be low when they are first introduced; this is a characteristic of all new technology. Nevertheless it will improve.
3. The cost is likely to be high; it has yet to benefit from the experience curve effect.
4. There is a psychological resistance to a new technology which replaces the expertise of the company and of individuals.
5. Large investments in manufacturing plant will be made obsolete; there is a desire to recover this investment by continuing to use it for as long as possible.

An explanation for some of these responses is that they are based on an analysis of the relative competitive strengths of the two technologies at one point in time. They ignore the dynamics of the evolution. Focusing on the future potential for performance improvement and on the dynamics of the technology substitution process forces a recognition of the inherent superiority of the new, and an acceptance that it will eventually become dominant. Failure to recognize this or a refusal to take action is one reason why new technologies are so often exploited by companies, frequently new ventures, with no vested interest in the current technology.

The discussion so far pre-supposes that a new technology has been identified. This may not necessarily be the case. However, the approach of a natural limit should alert the technologist. There may indeed not be a new technology. History indicates that there usually is. This can be seen by an examination of a number of technologies which show a succession of S-curves. They are often contained within an 'envelope curve', itself of S-shape. Ayres [8] has plotted the efficiency of the external combustion engine from the time of Savery (1698) and Newcommen (1712) to the high pressure turbines of the present day, spanning seven different technological approaches. In every case, he observes the same pattern. For example, between 1820 and 1850, the efficiency of the Cornish type engine rose rapidly from less than 5% to over 15%. Between 1850 and 1880, progress was slow until the triple expansion engine appeared on the scene, raising the efficiency to 22% by 1910, at which date the Parsons Turbine brought about another rapid rise. In the total period of 270 years, efficiencies have risen from virtually zero to over 50% and the 'envelope' curve describing them has closely followed the S-shape.

Precursor trends − curve matching
It is often possible to identify trends which precede the development in which one is interested. These can then be used as a cross-reference. They can take several forms:

1. Applications where cost and reliability are of lower importance.
2. The acquisition of knowledge before its practical application.
3. Adoption in a different geographical market.

Figure 8.8 exhibits a lag between the speed of combat and transport aircraft. The main reason for this is that the purchaser of military equipment is prepared to pay a high price and accept a relatively poor reliability if it meets his performance requirements, unlike the civil operator where these factors are much more important. Thus although the trends for both follow a similar curve there is a lag between them. A similar situation is often observed between specialist and mass applications. In the frequency of radio transmissions, for example, three stages of development can be traced − research, amateur and military, and full commercial applications. In all three a good correlation can be observed from the time of Marconi's first propagation experiment to the present-day laser, the time lag between research and commercial use being about thirty years with the amateur/military applications appearing about half way through this period.

The volume of patent registrations in a technology can also give prior warning of the growth of the technology based upon them. It is often observed that there is a typical period which elapses between the registration of a patent and the appearance in the market of products incorporating that knowledge. This has been noted by Koshi [10] for coating technology, Hakanson for the diffusion of special presses in paper-making [11] and Mackenzie for monoclonal antibodies [12].

Significant differences are also experienced in the diffusion of technology-based products in different markets. Frequently it is found that the growth in all markets follows the same shape but with a time lag; in many products the USA is the leading market. Twiss [13] has shown that the substitution of diesel for petrol engined cars in the UK and continental Europe has followed the same pattern in both markets with the UK market lagging by a constant ten years.

This discovery of a precursor relationship gives the forecaster additional confidence for he now has two 'fixes' for a point in the future, that obtained from extrapolating the curve for his own application, and a cross-reference from the precursor. It may also provide him with a useful clue to the date when a new substituting technology at present in use in the research laboratory, is likely to achieve commercial exploitation.

Attribute substitution
The desirability of a particular attribute to the customer may change over time. This may reflect changing attitudes or be a response to an event in his environment; an increase in fuel price, for example, may increase the desirability

of energy economy. In other cases the current level of performance may fully satisfy his needs without further progress. Such situations are likely to have important implications for the technological decision-maker who needs to address the following questions:

1. Should future investment or its emphasis be placed on a different attribute, i.e. improving the performance of parameter A or B rather than on C where it has been focused in the past?
2. Are there any environmental trends or events which may change the consumer preference between attributes?
3. What are the appropriate tradeoffs between parameters which are mutually interdependent?

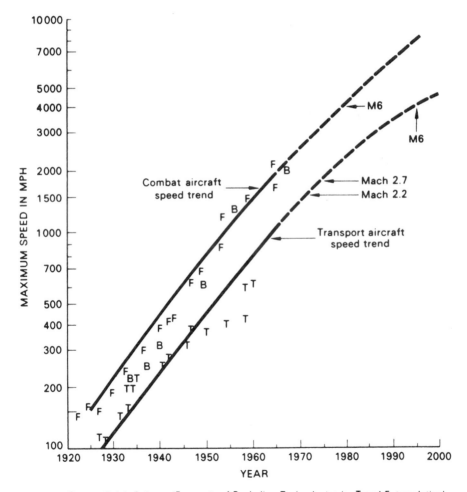

Source: Ralph C. Lenz, 'Forecasts of Exploding Technologies by Trend Extrapolation'

8.8 Speed trends of combat v transport aircraft, showing lead trend effect — what went wrong with the forecast?

The S-curves for three parameters relating to different attributes of the same product are illustrated in Fig. 8.9. It shows that parameter A is approaching its limit, implying that in the past effort has been concentrated on improving the performance of that parameter. The technological momentum this engenders together with the expertise and culture of the laboratory is likely to ensure that there is a natural tendency to continue along the same path. It should be noted, however, that both time and resources for an incremental improvement are increasing significantly. Furthermore, the current performance level may be sufficient to satisfy the needs of the customers; sometimes they may be unable to detect any slight improvement.

For a given technological investment consumer satisfaction might be enhanced

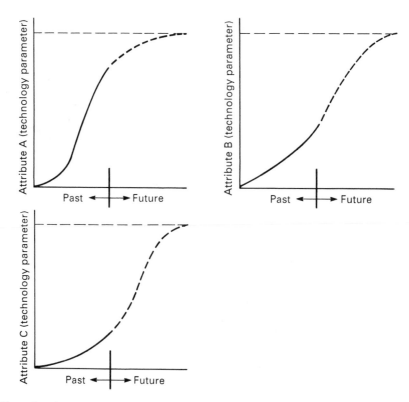

Note: Past investment has been focused on improving the performance of Attribute A.

Questions: 1. Should R & D effort be transferred from A to B or C?
2. Are there tradeoffs between A, B and C, i.e. will an improved performance in one reduce the performance in the others?
3. When should the change of focus be initiated?

8.9 Attribute substitution

by a greater improvement from another attribute. The time for a major realignment in the technological focus may have arrived. This can be illustrated by two examples. Traditionally the most desirable attribute required by a watch owner was accuracy in which the Swiss watch industry based upon precision instrument-making was dominant. The arrival of the highly accurate electronic watch made that attribute irrelevant; all watches are now accurate. The competitive advantage of the Swiss watch industry had been destroyed. Belatedly, with the introduction of the Swatch this was recognized by concentrating on another attribute, visual appeal. Similarly in domestic disinfectants technical performance was related to the microbial kill rate which was not only high but impossible for the user to assess. A competitive advantage was gained by the first manufacturer to concentrate on two other attributes, colour and perfume. In both these cases the market became less concerned with the purely technological attributes of the product.

The dangers of extrapolating past trends for a parameter without considering the total situation are illustrated by the forecasts for the speed of both civil and military aircraft made in the mid-1960s (Fig. 8.8). With the benefit of hindsight we know that these were poor forecasts, but why? A well established trend for the speed of both types of aircraft through a number of technologies can be seen for the period 1920 to 1965. An extrapolation during this period would have yielded a satisfactory forecast. However, for military aircraft, a technological development, air-to-air missiles, reduced the importance of speed. For civil airliners the reasons were different. It was noted earlier that the overall attribute, cost per passenger kilometre, was satisfied equally by both piston and jet engined aricraft. Thus whilst the overall market attribute remained unchanged it became more productive to satisfy it with a different emphasis.

Product substitution
Relatively few innovations introduce an entirely new capability into society. Where this does happen it is extremely difficult to make a realistic forecast of the eventual size of the market or its rate of growth. For example, the electronic calculator, whilst substituting for the abacus, log table and slide rule, achieved a market far in excess of that of all the earlier products. Such radical innovation is relatively rare and in most cases new technology manifests itself as a substitution for an existing product in one of two ways:

1. A complete substitution for the existing technology over a period of time. This almost invariably occurs in process innovations (e.g. basic oxygen for open-hearth steel, float for plate glass) where eventually the old technology disappears completely except perhaps for a very small residual amount.
2. A market segmentation. In such cases the new technology segments what was previously a homogeneous market. Part of the market will remain loyal to the earlier technology over a long period (e.g. wool and synthetic fibres, electric razors and wet shaving) whereas the other segment will eventually be completely replaced by the new technology. This makes the forecaster's

(a) Technology substitution curves

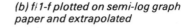

(b) f/1-f plotted on semi-log graph paper and extrapolated

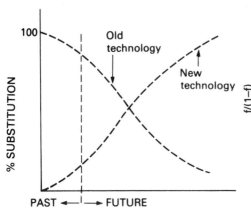

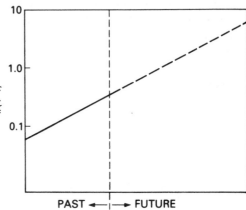

(c) New technology forecast as a % of the total derived from (b)

(d) Discontinuity in rate of substitution as a consequence of an economic, social or political change (e.g. the 1973 oil crisis)

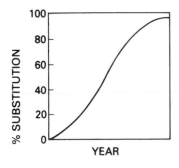

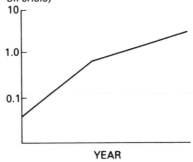

8.10 Product substitution

task more difficult since he must not only forecast the rate of substitution but also the size of the market segment to which the substitution applies. Furthermore, fashion and other new technologies may cause later changes in the relative sizes of the segments, for example a return in popularity of cotton at the expense of synthetic fibres has been occurring.

Studies of technological diffusion [14, 15, 16] covering a large number of substitutions show that they follow a pattern of a growing S-curve and a decaying curve (Fig. 8.10(a)). Fisher and Pry [17] showed that when the proportion of the new product reaches about 5% the dynamics of the substitution is likely to be sufficiently well established for the remainder of the curve to be forecast with reasonable confidence. This has provided the basis for one of the most useful techniques for the industrial technology forecaster.

The technique lends itself to a graphical representation for the mathematics of the curve are such that when the ratio of the proportion of the new technology (f) to the old ($1 - f$) is plotted on semi-logarithmic graph paper one obtains a straight line that can then be extrapolated (Fig. 8.10(b)). From this extrapolation it is possible to calculate the 'f' for future periods and plot on linear graph paper (Fig. 8.10(c)). It is then necessary to forecast the size of the total market (T) from which the size of the market for the new technology can be calculated ($f \times T$).

It must be remembered that the substitution curves are a model of human behaviour and do not represent a scientific principle; it is surprising how often one does observe a high correlation to a straight line in the semi-log plot. Examples of this are the substitution of radial ply tyres for cross ply in the UK (34% in 1969 to 93% in 1981) and at an earlier stage the substitution of diesel for petrol engines in the motor car in both the USA and Europe. Discontinuities can however, occur; Twiss (see [13]) has studied a number of substitutions over the period 1965–90 in which changes of slope occurred in the mid-1970s (Fig. 8.10(d)). Examples range from the substitution of colour for monochrome TV in the UK to the substitution of synthetic for cellulosic man-made fibres. These are remarkable in that they show a return to a straight-line relationship after a rapid change of slope. Once again it must be stressed that the forecaster can only extrapolate in the absence of new factors for which he should remain constantly on the lookout.

The diesel for petrol engined car substitution curves also exhibit two discontinuities. In the USA the sales of diesel cars plummeted in 1984 due to the poor reliability of diesel engines manufactured by General Motors and have yet to recover. In Germany, however, there was an accurate straight line correlation (on semi-log paper) from the mid-1960s until the late 1980s before a rapid fall occurred due to environmental concern with carbon particles in diesel exhausts. These examples show that no forecast can be immune from unforeseeable events, however well it has been conducted.

The experience (or learning) curve

The unit cost of production for a product falls as the cumulative number of items manufactured increases. There are a number of reasons for this: modification of the product design, use of cheaper materials, value engineering, improved manufacturing methods and technology, economies of scale and better management. Although it would be impossible to quantify the contribution each of these factors makes individually, it has been found in practice that in aggregate they result in a steady fall when the unit cost is plotted against the cumulative volume which has been produced. (Fig. 8.11(a) and (b)).

This phenomenon, first noted in the production of the Dakota aircraft during the war, is characteristic of both products and services. Typically the slope of the curve (Fig. 8.11(b)) is 70–80%, implying that the unit cost of production falls to 0.7 to 0.8 of its previous level every time the cumulative volume doubles.

The actual slope depends upon the characteristics of the product and the industry. This is another example of an observed relationship which can be used for forecasting future performance even when it is impossible to quantify the individual factors to obtain a causal model.

Consideration of the experience curve reveals a number of features of importance for technical planning.

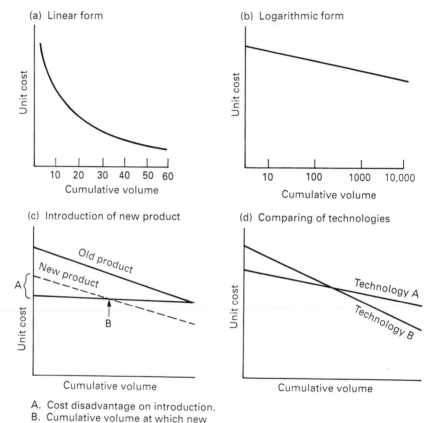

(a) Linear form

(b) Logarithmic form

(c) Introduction of new product

(d) Comparing of technologies

A. Cost disadvantage on introduction.
B. Cumulative volume at which new product becomes competitive.

8.11 The experience (or learning) curve

The high initial unit cost

The unit cost of a new product on introduction will be considerably higher than what will be achieved as volume builds up. If it is to be competitive with the product it replaces it must (Fig. 8.11(c)):

1. Incorporate features for which the customer will be prepared to pay a higher price.
   Or:
2. Have a substantially lower price, implying a radical technological improvement, such that its cost on introduction is equivalent or lower to

the current cost of the old product which has the benefit of the experience curve effect.

Or:

3. Be sold at a price below the production cost, sometimes referred to as 'experience curve pricing', until such time as the cumulative volume reduces the cost to that for the old product. Thereafter it will be more profitable. It should be noted that this involves additional market-related costs which must be added to the technical development costs when evaluating its desirability.

The rate of unit cost reduction

If the experience curve effect is ignored innovative proposals may be rejected because they are not evaluated on a like-for-like basis. Comparing the cost at introduction with a lower current cost for the existing product is not valid. The true basis for comparison must be their relative costs at equivalent stages in the cycle.

The competitive advantage of high volume

The high volume producer has a significant cost advantage. This will be associated with a high market share which is unlikely to be eroded by a small volume producer unless he can demonstrate a substantial performance improvement. Furthermore, this advantage must be maintained, an unlikely outcome unless the market leader can be excluded from introducing a similar product, for example by patent protection. Unless this can be done the advantage gained is likely to be short lived. This explains the continued dominance of IBM in mainframe computers and Boeing in wide-bodied jet aircraft.

A more desirable strategy for the new entrant or the low volume producer is to aim for a high market share of a niche or relatively small market segment. This removes many of the advantages of the leader who may also find the small size of the market unattractive.

The choice of technology

All technologies will not necessarily exhibit the same slope. Thus a technology which has a high unit cost at low volume relative to a competitive technology may fall to a lower unit cost at high volume (Fig. 8.11(d)). Amendola [18] has shown, for example, that for some automotive panels steel is cheaper than synthetic materials at high production volumes whereas the converse is true for low volumes.

Delphi

The opinion of experts can give important insights into the future, particularly in the identification of potential innovations likely to disturb the path of progress away from the extrapolated trend. Traditionally, expert opinion has been brought to bear through the medium of committee meetings. The Delphi technique was developed by Helmer at the Rand Corporation, to overcome the weaknesses of the committee by using the individual judgments of a panel of experts working systematically and in combination, divorced from the distortions introduced by their personalities.

The committee method suffers from a variety of shortcomings. In the first place geographical dispersion and the full diaries of prominent experts severely limit the membership and opportunities to bring them together. Once assembled, the committee process may not lead to a conclusion representing the unbiased views of all its members. Some people who are persuasive or articulate in discussion have a greater influence because of these characteristics rather than the strength of their case.

Other members may obtain a better hearing because of a position of authority or a high scientific reputation. Furthermore, there is a natural reluctance to change publicly a view which has previously been expressed strongly.

Another major weakness of the committee is the 'band-wagon' effect produced by the disinclination of an individual to disagree with a majority view, in spite of his own judgment. This phenomenon is illustrated with great force in the work of the psychologist Asch [19]. In a series of experiments, all but one of a group of students were briefed to support an erroneous view unbeknown to the remaining member. He described one of his experiments in which the subjects were shown two cards, one bearing a standard line, and the other three lines of which only one was the same length as the standard. When asked to select which of these three was the same length as the standard, Asch found that, 'whereas in ordinary circumstances, individuals matching the lines will make mistakes less that 1% of the time, under group pressure, the minority subjects swing to acceptance of the misleading majority's wrong judgment in 36.8% of the selections.' When such distortions can be created in a straightforward situation, one can appreciate the possible influence of this effect where there is genuine uncertainty as is inevitable when considering the future. Nevertheless, the minority view might well be valid and its expression should not be suppressed.

Delphi attempts to eliminate these problems by using a questionnaire technique circulated to a panel of experts who are not aware of the identity of their fellow members. The procedure is as follows:

1. *Round 1*. Circulate the questionnaire by post to the panel.
2. *Round 2*. After analysis of the Round 1 replies, recirculate, stating the median and interquartile range of the replies. Respondents are asked to reconsider their answers and those whose replies fall outside the interquartile range are invited to state their reasons – these may result from lack of knowledge, or more importantly, from some specialist information unknown to the other members.
3. *Round 3*. Analysis of replies from Round 2 are recirculated, together with the reasons proffered in support of the extreme positions, in the light of which the panel members are asked once more to reconsider their replies.
4. Further rounds for additional clarification may be employed if thought necessary.

Selection of the panel members is a task of the utmost importance for the value of the forecast is a function of the calibre and expertise of the individual contributors as well as the appropriateness and comprehensiveness of the areas of knowledge they represent. Questionnaire formulation also requires considerable skill to ensure the right questions are asked and that they are framed in specific, quantified and unambiguous terms.

Delphi remains one of the most widely used forecasting methods (Linstone and Turoff [20]) but is not without its critics (Sackman [21]). Many studies have been poorly conducted, ambiguous questions being a common failing. It should not be used as an easy option where more rigorous techniques could be applied. However, well-conducted Delphi studies will continue to be carried out particularly for:

1. Long-range forecasts.
2. Probability estimates where more objective methods do not exist.
3. The occasional benefit of new insights.
4. Time and quantitative forecasts only where other methods cannot be used.

But how accurate are the resulting forecasts? There are some indications that there may be a tendency to err in an optimistic direction in the short term due to an underestimation of development times. By contrast, long-term forecasts may well be pessimistic because of the mind's inability to appreciate fully the effects of exponential growth.

Rand Corporation has, however, conducted experiments to validate the methodology. These show clearly that when experts have been asked about current phenomena where it is possible to establish a factual answer, consensus grows as the study progresses, and moves towards the correct answer. This suggests possible applications outside TF such as in cost estimation and R & D project selection.

Scenarios

Scenario writing has become most widely known through the work of American 'think tanks'. The scenarios describe a possible future situation based upon a wide-ranging environmental analysis. Frequently several scenarios or alternative futures are prepared supported by detailed research using a wide variety of TF techniques. Scenario writing is based upon the recognition that it is not always possible to choose between alternative sets of assumptions.

In recent years interest in scenario writing has increased at both the national and industrial level. Energy forecasting is a field where this approach has been used extensively. A number of practical techniques for industrial scenario writing have been developed [7, 22] to enable the consideration of the mutual interactions of a wide range of environmental factors both upon themselves and upon an organization's strategic objectives. Thus this is an approach which extends beyond technological planning and enables top management to review their strategic assumptions and the consequences flowing from them. Scenarios have been used as the basis for their corporate planning by Shell since the early 1970s due to a wide range of political and economic uncertainties inherent in the oil industry.

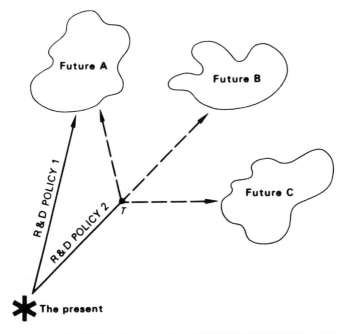

Futures A, B and C are three scenarios of alternative futures . The forecast may result in R & D action as follows:

POLICY 1. Future A is considered so probable that decisions are made assuming the forecast is correct.

POLICY 2. A mimimum risk policy permitting an advance to *T*, without precluding any of the scenarios. At time *T* a decision cannot be deferred any longer. In this way the decision has been delayed to the latest possible date.

8.12 A minimum risk policy for possible alternative futures

The use of scenarios in decision-making is illustrated diagrammatically in Fig. 8.12. Three alternative futures have been identified: A, B and C. Although it is possible to attach probabilities to the likelihood of occurrence of each of the alternatives, they are not likely to be very reliable. It may, of course, emerge that all the evidence points in one direction with a high probability of the future being similar to that described by Scenario A. In this case, the long-term R & D programmes (Policy 1) could be based on this assumption. Policy 1 might, however, be such that it has no relevance to Futures B and C. In the more likely case when it is not possible to choose clearly between A, B or C, a programme (Policy 2) which keeps the options open is likely to be the best compromise. At a later date the programme can be reorientated in the light of future events. Thus a minimum risk strategy can be adopted as a conscious act of policy.

Most writers stress the interaction between the decisions resulting from forecasts and the determination of the future. If Scenario A is thought to be likely then following Policy A as a result of the forecast could make it a self-fulfilling prophecy. Such freedom to shape the future is open to few companies. But one can see how scenarios could be used to shape government policy. If, for example, Scenario C described a highly undesirable future for mankind, then, given the correct political conditions, government policies could ensure that the path leading to it is sealed off.

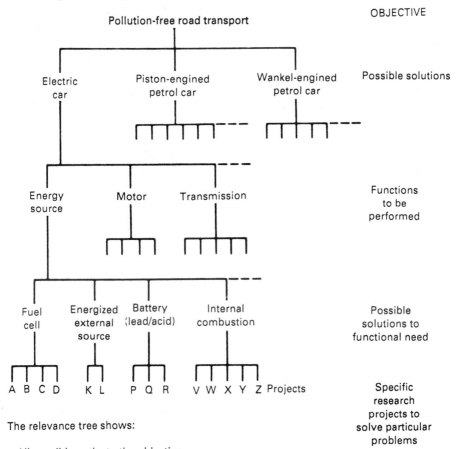

The relevance tree shows:

- All possible paths to the objective
- Forecasts of costs, durations and associated probabilities for each element

It assists in:

- Establishing the feasibility of the objective
- Selecting the optimum programme
- Estimating overall cost and duration
- Scheduling the detailed R & D programme

8.13 Relevance tree

Relevance trees

The purpose of a relevance tree is to determine and evaluate systematically the alternative paths by which a normative objective or mission could be achieved (Fig. 8.13).

The objective (O) is the starting point – let us say a pollution-free road transport system. At the next level, the problem can be broken down into alternative solution concepts A, B, C, etc. (e.g. electric car, Wankel engine, 'clean' petrol piston engine, etc.), or by functions to be performed (e.g. for a moon landing there would be launch, mid-course flight and control, lunar landing, lunar take-off, earth re-entry). For a particular solution a variety of systems will be necessary (e.g. fuel systems, combustion, and exhaust), each of which in turn may be satisfied in a variety of ways involving alternative subsystems (e.g. mechanical, electrical or catalytic exhaust cleaners).

Thus, starting from the desired end result, it is possible to examine exhaustively all the alternative paths by which it can be reached, working backwards through a hierarchy ending with detailed and limited R & D project objectives. The next stage is to investigate each step in greater depth including feasibility, resources required, probability of success and timescale. This involves the employment of other TF techniques such as Delphi and trend extrapolation.

Relevance trees can be useful to the R & D manager in:

1. Establishing the feasibility of a technological mission. If no feasible path can be found, the mission cannot be achieved with the technological capability of the present or the foreseeable future.
2. Determining the optimum R & D programme, i.e. the paths to be followed up the hierarchy. This could vary with the programme objective. Early achievement of the objective may call for a different programme (i.e. path) from that for minimum cost.
3. Selecting and planning the initiation of detailed research projects.
4. Establishing performance objectives for the constituent parts of an R & D programme.
5. Identifying the need for detailed technology forecasts of the contributing technologies where their performance within a given timescale is important.

Cross-impact analysis

The techniques we have discussed so far have centred on forecasting the future behaviour of specific attributes or likely events in isolation, although we have seen that in fact many of them are interdependent not only within the technological sphere but also between technology and other environmental factors. Gordon of the Institute of the Future, classifies these interconnective modes as *enhancing* (enabling or provoking) or *inhibiting* (denigrating or antagonistic). Thus our forecasts for automotive anti-pollution legislation enhance the probability of the development of catalytic exhaust cleaners, but the development of non-catalytic exhaust cleaners would inhibit the probability of developments of new processes for purifying platinum, the future demand for which is closely related to the potential demand for catalysts.

The cross-impact technique enables the systematic examination of the relationships between a range of factors. It usually takes the form of a matrix in which the same factors are listed on both the horizontal and vertical axes. Their mutual impacts are then assessed and entered into the matrix in respect of both the direction of the influence and its magnitude. Some of these relationships might be quantitative (e.g. GNP *v* energy consumption) but most are likely to represent the judgments of the panel conducting the analysis. It is particularly valuable in that it enables the consideration of economic, social and political as well as technological developments. A description of the procedure is given in Twiss [see [7] earlier].

In essence the procedure is analogous to the development of an influence model with the number of factors reduced to what can be handled conveniently.

The main uses of a cross-impact analysis are:

1. To establish sets of coherent assumptions for the generation of alternative scenarios.
2. The modification of probabilities assigned in simple forecasts to take account of impacting events.

Organization for technology forecasting

For the majority of companies, the employment of more than two or three full-time technological forecasters is the exception. In most of the companies where the techniques have been studied the involvement has been largely confined to a part-time activity of a few interested staff in the R & D department.

The best person to make a forecast is the manager who has to make a decision which is future-related. Usually, however, he will need assistance in two areas:

1. The provision of data.
2. The application of TF techniques.

Much of the data required for TF is of the type which is needed in several areas of the company (e.g. economic and statistical data, market research) and is most conveniently collected and stored centrally within the organization, perhaps in the economics or corporate planning section. Other data which is of a specialist nature, of interest only to the departments preparing the particular forecast, needs to be collected at departmental level. The value of such data storage grows with time since early efforts at TF are frequently frustrated by the absence of information. Thus the first and critical step in adopting TF is to lay the foundation for future forecasting by organizing the collection and recording of data.

Few managers have been trained in TF and it is impracticable to provide the training for all those who might be called upon to make forecasts. There is thus need for consultancy advice from a small staff who are well versed in the techniques and can advise the manager in the preparation of his forecasts and warn him of the many pitfalls. Nevertheless it is better that the individual manager makes his own forecast, with assistance, rather than the removal of forecasting to a central service department remote from the problem area.

Since all technological decisions relate to the future the managers taking those decisions must use any techniques which can aid them. TF enables a systematic analysis of what has happened in the past and provides a framework for assessing future trends. It is, therefore, an important tool for all technical managers, particularly in so far as it indicates where a change in technological strategy can provide a competitive advantage.

The full application of some of the more sophisticated techniques requires training, experience and assistance. However, this is a counsel of perfection since many organizations are not prepared to make the necessary investment. The onus is then on the individual manager to do the best he can for himself. This is not so difficult as it may seem. A working competence in many of these techniques can be obtained without undue difficulty. As experience is gained so will the quality of the forecasts improve.

Forecasting is not a one-off exercise. It must be a continuous activity. With time the systematic collection of data will provide an improved basis for conducting the forecasts. The first attempts are likely to be tentative and there may be little confidence in them. Nevertheless when it is possible at a later date to compare the forecast with what actually happened a great deal can be learnt. It is quite likely that it will be found that if decisions had been based upon those forecasts, in spite of their deficiencies, the outcome would have been more satisfactory. If a significant difference is revealed it might indicate where the forecasting procedure might be improved; it is all part of the learning process. In some cases, of course, the forecast might have been invalidated by an event which no forecasting activity could have identified. It must be stressed again that forecasting can never ensure that the right decision is taken; the best that can be hoped for is that decisions based on forecasts will be better than those taken in their absence.

Current status of technology forecasting

Most of the initiatives in TF originally came from the USA where several major companies invested heavily in it. There is some doubt whether the results justified the substantial investments.

In Europe the rate of adoption has been much slower. The fuel crises of recent years have given a stimulus to industrial concern with the future and a growth in forecasting activity. It is possible to identify two levels of support; the first is concerned primarily with R & D planning decisions whereas the second focuses on corporate strategic issues. The latter is often referred to as 'future studies' rather than as forecasting. There is also a recognition that the future is more uncertain now than it was in the recent past and, furthermore, it is more likely to experience discontinuities. Thus we are witnessing a growth in the use of scenarios.

Many of the techniques lend themselves to programming on microcomputers (e.g. relevance trees, cross impact). This enhanced ability to deal with large quantities of data and carry out sensitivity analyses easily, greatly adds to the forecasting capability provided the forecaster remembers the need to remain mentally alert without being blinded by the sophistication of the techniques.

A major cause for the low rate of acceptance is likely to remain the reluctance to invest the large resources, which the literature suggests are necessary, in an untried technique. This has not encouraged more modest attempts. Nevertheless, as the pace of technological progress continues to increase, so will the need to forecast. Thus we may expect to see a growth in the use of TF. But this growth is likely to be hindered by exaggerated claims for what it can achieve.

Another factor contributing to the relatively low level of TF activity in the West is the short-term orientation of many companies [23]. It is notable that there are few recent references to TF in the technology management literature. However, by contrast a study conducted by the author in Japan [24] showed that the long-term approach in that country is supported by a high level of TF activity using techniques which were developed in the West. This applies at both the governmental and company level as exemplified by:

1. The Japanese Government Science and Technology Agency conducts a macro-technology forecast at five-year intervals. This takes the form of a 30-year TF using a Delphi study involving 1,500 participants drawn from industry, government and academia.
2. Hitachi has an 80-member 'think tank' as a subsidiary company of which the Hitachi president is also the president.
3. Matsushita has 10 full-time corporate forecasters and a further 20 in R & D.
4. Mitsubishi Electric has 50 trained technology forecasters.

Summary

Since the benefits from R & D decisions are gained in the future, it is incumbent upon the R & D manager to satisfy himself, so far as is possible, that the results of his investments are relevant to the market needs and the competitive technologies at the time they reach fruition. Thus all R & D decision-makers must take a conscious view of the future. Forecasts are needed which take full account of the information available and the techniques of TF. It was seen, however, that the effort devoted to TF should be related to the characteristics of the industry, the company, and the decisions to be made, rather than to the size of the company.

Technology forecasting cannot enable the decision-maker to predict the future with certainty. But it can assist him in refining his judgments. The value of the forecasts was seen to be highly dependent upon the quality of the informational inputs to the forecasting process and the calibre of the minds applied to it. Sophisticated forecasting techniques can only be aids to this process, and care should be taken to guard against TF absorbing greater resources than can be justified in economic terms.

The principles underlying technology forecasting were discussed and brief descriptions given of how they are applied in some of the most commonly used techniques – trend extrapolation, precursor trend curve matching, substitution analysis, Delphi, scenarios, relevance trees and cross-impact analysis.

Whilst the practical application of TF is still limited, it was concluded that its use is likely to become more widespread once a better understanding is gained of what it can contribute to R & D decision-making and, perhaps more important, what it cannot be expected to do. No R & D manager can afford to ignore TF, but enthusiasm for the techniques must be tempered by the realization that they alone will never remove completely the uncertainties inseparable from any consideration of the future.

References

1. Drucker, P. F. *Technology, Management and Society*, Heinemann, 1970.
2. Twiss, B. C. 'Economic Perspectives of Technological Progress: New Dimensions For Forecasting Technology', *Futures,* **8**, No. 1, Feb. 1976.
3. Foster, R. *Innovation: The Attacker's Advantage*, Macmillan, 1986.
4. Isenson, R. S. 'Technological Forecasting Lessons from Project Hindsight', in Bright, J. R. (ed.), *Technological Forecasting for Industry and Government; Methods and Applications,* Prentice-Hall, 1968.
5. Jones, H. and Twiss, B. C. *Forecasting Technology for Planning Decisions,* Macmillan, 1978.
6. Martino, J. P. *Technological Forecasting For Decision Making*, 2nd edn, Elsevier, 1983.
7 Twiss, B. C. *Forecasting for Technical Decisions*, Peter Peregrinus, 1992.
8. Ayres, R. U. 'Envelope Curve Forecasting', *Technological Forecasting for Industry and Government*, Bright, J. R. (ed.), Prentice-Hall, 1968.
9. Lenz, R. C. 'Forecasts of Exploding Technologies by Trend Extrapolation', in *Technological Forecasting for Industry and Government* – see reference 4.
10. Koshi, T. 'Trend and Development Forecasts of Diversified Coating Technology, *Patents and Licensing*, Aug. 1974.
11. Hakanson, S. 'Special Presses in Paper Making', in *The Diffusion of New Industrial Processes*, Nasbeth, L. and Ray, G. F. (eds), Cambridge University Press, 1974.
12. Mackenzie, M. 'The Commercial Application of Scientific Discovery: The Case of the Hybridomia Technique', *Research Policy*, **17**, No. 3, June 1988.

13. Twiss, B. C. 'Forecasting Market Size and Market Growth Rates for New Products', *Product Innovation Management*, **1**, No. 1, Jan. 1984.
14. Nasbeth, L. and Ray, G. F. *The Diffusion of New Industrial Processes*, Cambridge University Press, 1974.
15. Ray, G. F. *The Diffusion of Mature Technologies*, Cambridge University Press, 1984.
16. Linstone, H. A. and Sahal, D. *Technological Substitution*, Elsevier, 1976.
17. Fisher, J. C. and Pry, R. H. 'A Simple Substitution Model of Technological Change', *Technological Forecasting and Social Change*, **3**, 1971.
18. Amendola, G. 'Diffusion of Synthetic Materials in the Automobile Industry', *Research Policy*, **19**, No. 10, Dec. 1990.
19. Asch, S. E. 'Opinions and Social Pressures', *Scientific American*, November 1955. Reprinted in: *Readings in Managerial Psychology*, Leavitt, H. J. and Pondy, L. R. (eds), University of Chicago Press, 1964.
20. Linstone, H. A. and Turoff, M. *The Delphi Method: Techniques and Applications*, Addison-Wesley, 1975.
21. Sackman, H. *Delphi Critique*, Lexington Books, 1975.
22. MacNulty, C. A. R. 'Scenario Development For Corporate Planning', *Futures*, **9**, No. 2, April 1977.
23. Twiss, B. C. *Non MSC Research and Initiatives in the Field of Management Development and Change*, Manpower Services Commission, 1985.
24. Goodridge, M. and Twiss, B. C. *Management Development and Technological Innovation in Japan*, Manpower Services Commission, 1986.

Additional references

Becker, R. H. and Spelt, Z. 'Putting the S-Curve Concepts to Work', *Research Management*, **XXVI**, No. 5, Sept.–Oct. 1983.

Boucher, W. I. (ed.), *A Study of the Future: An Agenda for Research*, National Science Foundation Report NSF/RA-770036, 1977.

Bright, J. R. *Practical Technology Forecasting: Concepts and Exercises*, Industrial Management Centre, 1978.

Fowles, J. (ed.), *Handbook of Futures Research*, Greenwood Press, 1978.

Jantsch, E. *Technological Forecasting in Perspective*, Organization for Economic Cooperation and Development (OECD), 1967.

Jantsch, E. *Technological Planning and Social Futures*, Cassell/Associated Business Programmes, 1972.

Linstone, H. A. and Simmons, W. H. *Futures Research: New Directions*, Addison-Wesley, 1977.

Sahal, D. *Patterns of Technological Innovation*, Addison-Wesley, 1981.

Twiss, B. C. *Social Forecasting for Company Planning*, Macmillan, 1982.

Index